ANDREA ROBERTO CODRIGNANI

Numerical representation of a pin-on-disc tribometer for the investigation of textured surfaces

SCHRIFTENREIHE DES INSTITUTS FÜR STRÖMUNGSMECHANIK
KARLSRUHER INSTITUT FÜR TECHNOLOGIE (KIT)

BAND 6

Eine Übersicht aller bisher in dieser Schriftenreihe erschienenen Bände finden Sie am Ende des Buchs.

Numerical representation of a pin-on-disc tribometer for the investigation of textured surfaces

BY

ANDREA ROBERTO CODRIGNANI

KIT Scientific Publishing

Dissertation, Karlsruher Institut für Technologie (KIT)
KIT-Fakultät für Maschinenbau

Tag der mündlichen Prüfung: 5. Oktober 2018
Referenten: Prof. Dr.-Ing. B. Frohnapfel, Prof. Dr. B. Bou-Saïd

Impressum

KIT Scientific Publishing

Karlsruher Institut für Technologie (KIT)
KIT Scientific Publishing
Straße am Forum 2
D-76131 Karlsruhe

KIT Scientific Publishing is a registered trademark
of Karlsruhe Institute of Technology.
Reprint using the book cover is not allowed.

www.ksp.kit.edu

Print on Demand 2019 – Gedruckt auf FSC-zertifiziertem Papier

ISSN 2199-8868
ISBN 978-3-7315-0874-8
DOI 10.5445/KSP/1000088256

Abstract

The growing demand for higher performance and lower environmental impact of tribological devices has led to the development of textured surfaces as efficient way to reduce friction losses and wear. Great part of the tribological research of such engineered surfaces relies on experimental set-ups based on pin-on-disc tribometers. The aim of the present work is the numerical representation of such kind of experimental set-ups in order to study the influence of textures in the hydrodynamic regime.

In the initial analysis of the mentioned set-up, particular attention is paid to the high-resolution representation of the macroscopic geometry of the contact. In this regard, the influence of the velocity gradient in the plane of rotation is investigated and compared to experimental studies. The role of inertia effects is then examined by means of the Navier-Stokes equation and the validity of the Reynolds equation is investigated for 3D dimples resulting in the extension of previous literature results. The effects of viscous dissipation are also considered in order to assess their influence on the pressure losses in the contact, which are found to scale with the square of the Sommerfeld number.

Finally, the influence of the operating conditions on the optimal shape parameter of a texture made by dimples is investigated by means of an extensive parametric study. As a result, the dimple depth is shown to have a higher impact on the texture performance than the diameter. Moreover, the dimple depth, which corresponds to the minimal friction coefficient, is found to scale with the square root of the Sommerfeld number in agreement with experimental investigations on the same tribometer.

Kurzfassung

Die wachsende Nachfrage nach höherer Leistung und geringerer Umweltbelastung durch tribologische Geräte hat zur Entwicklung strukturierter Oberflächen geführt, die Reibungsverluste und Verschleißphänomene reduzieren können. Ein großer Teil der tribologischen Forschung an solchen Oberflächen beruht auf experimentellen Untersuchungen, die auf Pin-on-Disc-Tribometern basieren. Ziel der vorliegenden Arbeit ist, die numerische Darstellung dieser Experimente und die Untersuchung des Einflusses von Texturen im hydrodynamischen Bereich.

Bei der ersten Analyse des Pin-on-Disc-Tribometers wird besonders auf die hochauflösende Darstellung der makroskopischen Geometrie des Kontaktes geachtet. In dieser Hinsicht wird der Einfluss des Geschwindigkeitsgradienten in der Rotationsebene untersucht und mit experimentellen Studien verglichen. Die Rolle von Trägheitseffekten wird anhand der Navier-Stokes-Gleichungen analysiert und die Gültigkeit der Reynolds-Gleichung für 3D-Dimples überprüft. Die Auswirkungen der viskosen Dissipation werden ebenfalls berücksichtigt, um deren Einfluss auf die Druckverluste im Kontaktbereich zu bestimmen, die mit dem Quadrat der Sommerfeldzahl skalieren.

Schließlich wird mittels einer umfangreichen Parameterstudie der Einfluss von Betriebsbedingungen auf die optimalen Formparameter einer durch Dimples strukturierten Oberfläche untersucht. Als Resultat dieser Untersuchung folgt, dass die Tiefe der Dimples einen höheren Einfluss auf die gesamte Kontaktleistung hat als der Durchmesser. Darüber hinaus wird gezeigt, dass die Tiefe der Dimples, die dem minimalen Reibungskoeffizienten entspricht, mit der Quadratwurzel der Sommerfeldzahl skaliert und damit in Übereinstimmung mit experimentellen Untersuchungen am gleichen Tribometer steht.

Table of Contents

1 Introduction

> Nature is the source of all true knowledge. She has her own logic, her own laws, she has no effect without cause nor invention without necessity.
>
> Leonardo da Vinci

The interaction between moving surfaces is a concept which is at the heart of innumerable phenomena in the nature. Whether in the joints of a human skeleton, in rolling elements of a jet turbine, or in crucial components on a space station, no natural events nor human-designed devices are exempt from the action of friction and wear. The intrinsic multidisciplinary character of these two physical phenomena aroused the interest of mankind since the ancient times and particularly regained great attention in Renaissance with the pioneering studies of Leonardo and subsequently during the industrial revolution [1, 2]. Nonetheless, the definition of the concept of tribology in terms of scientific discipline and economical implications originated first in the second half of the twentieth century thanks to the work of Jost [3]. By means of a careful analysis of the state of lubrication research in different industrialized countries, the Jost Report and many following independent studies estimated a potential annual saving ranging from 1% to 1.4% of a country's GDP [4–6]. Furthermore, such saving is deemed to be achievable with a very convenient return on investment ratio of 1/50, implying a saving of 50 $ for each dollar spent in research and development in the previous year [4, 7]. Particular emphasis is laid on the role of tribological progress in transportation, industrial and utilities sectors, where it has been estimated that up to 11% of the used energy can be saved by the application of the new developments in tribology

[8, 9]. In view of such relevant potential impact, the advance in the field of tribology is nowadays of utmost importance to address the economical and environmental challenge and also to cast light on a plenty of phenomena in nature which characterize our everyday life.

1.1 Motivation

In the last decades, the research on tribology has spread in numerous branches, ranging over multidisciplinary areas such as lubricants development [10, 11] surface coating [12, 13], or the optimization of automotive and industrial applications [14, 15]. Among these research fields, the enhancement of the tribological performance through the introduction of surface textures has drawn a considerable attention in the research community [16]. Surface texture nowadays represents an interesting technology for the reduction of friction and wear thanks to recent advancements in the laser surface texturing techniques (LST) which reduced the production costs and increased also the manufacturing precision [17].

However, an unanimous consensus in the research community has not been reached yet, for what concerns the underlying physical mechanisms and the possibility to obtain an optimal texture design which proves to be robust under different operating conditions [18]. In order to bridge this gap, a more and more increasing part of literature combines experimental and numerical works [16]. Among them, the present work represents the numerical counterpart of the three experimental studies on surface textures which were carried out with a pin-on-disc tribometer [19, 20, GMB+15]. For this reason, this introduction will first focus on the state of art of the experimental and numerical research carried out with pin-on-disc tribometers, leaving further extensions in the literature survey of each following chapters.

In the investigation of surface textures two main approaches are usually adopted. In the first case, the whole component is studied in the framework of the final application, while in the second one, the tribological contact is isolated and investigated in a simplified system. To the first category belong most of the studies on thrust bearings [21–25], journal bearings [26, 27], seals [28, 29] or piston rings [30–32]. This first approach has the advantage to allow the investigation of the impact of surface textures on the final performance of the considered device. On the other hand, the increased complexity of the set-up reduces the feasibility of extensive studies with different texture designs. Therefore, a consistent part of the experimental research on surface texture is carried out with tribometers in pin-on-disc configuration. This kind of experimental apparatus has the great advantage of reproducing a generic sliding contact in a simpler and controlled way, still giving the possibility to analyze the influence of a large set of design parameters. Moreover, it allows to test different texture designs more rapidly due to the easiness in changing the textured specimen. This feature can be also exploited in order to easily investigate the combined effects of textures on different materials, as done, for example, by Wakuda *et al.* using silicon nitride ceramic pin on hardened steel disc [33] or combining special surface treatment [34] or different steel alloys [35].

One of the first papers which investigated the effectiveness of surface textures with a pin-on-disc tribometer was presented by Kovalchenko *et al.*, who found that, if properly designed, surface textures can extend the hydrodynamic range for both high- and low-viscosity lubricants, leading to lower friction coefficients under the same operating conditions [36]. The experimental work by Braun *et al.* [19] comes to similar conclusions and also points out that the shift of the hydrodynamic range scales in good agreement with the operating conditions. The results of this experimental work will be further analyzed in chapter 4 of the present thesis. The change of lubrication regime induced by surface textures is further investigated by Zhang *et al.* under mixed and starved lubrication regime [37]. A similar approach based on pin-on-disc tribometers was also used by Scaraggi *et al.* in two works [38, 39], showing that friction reduction can be achieved by surface texturing in both mixed and lubrication regime.

The results of the above-mentioned studies confirm the different working principles which have been identified also through other experimental techniques and numerical analyses. These can be resumed as follows: firstly, in the boundary regime or under starvation conditions, textures are deemed to act as lubricant reservoirs [33, 40], hence providing lubricant to the contact and at the same time entrapping wear debris and reducing third-body abrasion [41, 42]. Secondly, thanks to the reduction of contact area, the probability of adhesive wear and the static friction may be diminished [43, 44]. Finally, the presence of texture can generate an additional hydrodynamic lift that can increase the load-carrying capacity in mixed and hydrodynamic regimes. More specifically, the hydrodynamic lift can be induced by inertial effects [45, 46], by an asymmetrical pressure distribution caused by cavitation [47] or by partial texturing [48, 49].

A part of these hydrodynamic effects is investigated also by means of joint experimental and numerical works based on pin-on-disc tribometers. For example, Ramesh *et al.* combine the experimental activity with a numerical model based on the Navier-Stokes equation which analyzes the increased pressure distribution over a dimpled surface due to nonlinear effects [50]. Their model is used to predict the friction coefficient as function of the texture shape, although the numerical part focuses only on a single texture element, because of the high computational cost of Navier-Stokes simulations.

Once assessed the positive influence which surface textures can have on the tribological performance, the focus can be moved to the optimization of different design parameters, such as the texture arrangement and its density. In this regard, Yu *et al.* integrate the pin-on-disc measurement with a numerical analysis based on the Reynolds equation in order to cast light on the influence of different texture arrangements [51]. In particular, they study the disposition of rows of dimples as a function of the shift between every line, showing that an optimum value can be found which minimizes the friction coefficient. However, cavitation phenomena are neglected and little information is provided about the mesh and the numerical framework, leaving some questions opened. For this reason, a similar experimental analysis was carried out by Schneider *et al.* in 2017, by varying the shift between consecutive dimple rows in a more systematic way [20]. A numerical analysis based on this experimental results is performed in this thesis and its results are shown in section 4.5.

The pin-on-disc set-up is used also for the investigation of different texture shapes, as done for example by Morris *et al.* [52]. In this respect, particular attention has to be paid to the precise investigation of various dimple shapes without changing other sensitive parameters, such as texture density and dimple depth, as underlined by Etsion in [18]. Some pin-on-disc tribometers can also employ rotating discs made out of transparent materials, in order to allow the identification of cavitation zones in the macro contact as well as inside the texture [53], or in order to measure the size of the contact area [54].

As a matter of fact, great part of the experimental activity on textured surfaces is carried out with pin-on-disc tribometers. For this reason, it has become important to investigate the influence of certain characteristic parameters of such an experimental device also from the numerical point of view. In this regard, the work of Greiner *et al.* [GMB+15] analyzes the impact of the distance between the specimen and the center of rotation on the friction coefficient with and without textures. The experimental evidence, which shows lower friction at higher distance, is also confirmed numerically through CFD simulations, whose details are presented in the present work in section 2.3.1. Other works in literature investigate the pin-on-disc tribometer without texture with the aim to give useful insights on the physical peculiarities which characterize it, such as dry friction and wear phenomena [55, 56] and viscous heating [57].

More generally, for what concerns the numerical representation of pin-on-disc tribometers, most of the works in literature consider two main hypotheses, namely that the specimen is untextured and that its shape is assumed to be spherical. This geometry is also known as ball-on-disc and is typically studied in the framework of elastic hydrodynamics (EHD) due to the possibility to obtain an analytical solution with such a contact shape [58, 59]. Numerical solutions of the EHD problem were proposed, among others, by Lubrecht and co-workers [60, 61] through multigrid methods, by Doki-Thonon *et al.* for non-Newtonian lubricants [62] and by Habchi *et al.* [63] by employing a finite element solver for the contact mechanics coupled with the Reynolds equation. Thermodynamics can also be added to elastohydrodynamic models (TEHL) for both Newtonian [64, 65] and non-Newtonian fluids [66] through the coupling of the Reynolds equation with the energy one. Alternatively to the Reynolds equation, Bruyere *et al.* solved the compressible Navier-Stokes equations in the case of sliding line [67], providing a physical description also across the gap height.

However, the computational power required by the aforementioned works limited the analysis to untextured surfaces or even to 1D contact line. This issue becomes even more evident when cavitation phenomena are considered, because of the relatively low convergence speed of the usual iterative schemes, such as, for example, the $p-\theta$ algorithm by Elrod and Adams [68]. In this regard, a big step forward has recently been done by Woloszynski *et al.* in 2015 by developing a mass-conserving cavitation scheme based on the so-called Fischer-Burmeister-Newton-Schur (FBNS) algorithm [69]. This algorithm, which solves the cavitation problem in the framework of an unconstrained optimization, presents a convergence speed which is up to two orders of magnitude faster than previous methods.

As a conclusion to this literature survey, the relevance of the pin-on-disc tribometer for the experimental investigation of surface textures is highlighted and the numerical works based on this set-up are introduced. To the author's knowledge no previous work addressed the numerical representation of the full-size specimen of pin-on-disc tribometer with mass-conserving cavitation algorithm in order to investigate the impact of surface textures.

1.2 Objectives and procedure

The objective of this thesis is the numerical representation of a pin-on-disc tribometer in order to study the influence of surface textures in tribological contacts. The path to achieve this goal can be divided in two major parts. The first one concerns the analysis of the physical phenomena which characterize the hydrodynamic regime in order to determine which of them needs to be modeled and which spatial scales needs to be considered for the analysis of surface texture in this lubrication regime.

Therefore, the large-scale effects, namely those induced by the macro-geometry of the tribometer, are addressed by means of direct numerical simulations with the Navier-Stokes equation. In this framework, both the role of different set-up configurations and inertia effects are analyzed. The latter are further analyzed at smaller scales, namely on a single texture element, through the comparison between the Navier-Stokes and the Reynolds equations.

Subsequently, the impact of thermodynamic effects is discussed, focusing on the influence of viscus dissipation on the pressure and temperature distribution in the lubricant film. For this analysis both the Navier-Stokes and the Reynolds equations are coupled with the energy equation and particular attention is paid to the efficient implementation of the thermal-Reynolds solver. In this way, a comprehensive parametric study can be carried out under different operating conditions in order to assess the scaling of viscous dissipation effects.

In the second part of this work, the high-resolution representation of the experimental set-up is used as basis for an extensive and systematic study of textured surfaces. Thanks to the prior analysis about the modeling of the hydrodynamic effects, the incompressible Reynolds equation with mass-conserving cavitation is chosen as the best compromise to grant good accuracy and high computational performance. The pin geometry is firstly studied in the framework of a sensitivity analysis, focusing on the part of the domain in which the presence of a texture improves the tribological performance. Subsequently, optimal texture parameters, both in terms of load-carrying capacity and friction reduction, are studied with respect to the operating conditions; particularly viscosity, velocity and gap height. The resulting scalings and trends are finally compared to the previous experimental analysis carried out on the same set-up and generalized with respect to the Sommerfeld number.

2 Fundamentals

This chapter introduces the tribometer set-ups and the numerical models which are used to describe the physical phenomena in the hydrodynamic regime. Moreover, a preliminary analysis is discussed which deals with the impact of important set-up parameters, such as the sliding radius and relevant flow aspects, such as inertial effects.

Tribometers can actually operate under different lubrication regimes, which are consequently dominated by different physical phenomena. A conceptual representation of the different regimes is depicted by the Stribeck curve [70] as shown in figure 2.1. The abscissa in the figure represents the dimensionless number, also referred to as the Hersey number, which is defined as follows:

$$\mathcal{H} = \frac{\mu\Omega}{W} \tag{2.1}$$

where μ is the dynamic viscosity, Ω the relative angular velocity and W the average load applied.

Low Hersey numbers usually correspond to a small gap height and identify the boundary lubrication regime, in which there is a significant asperity contact and a lack of lubricant, leading to high friction coefficients (the latter is defined in followings in equation 2.2). As the Hersey number increases, a noticeable decrease of the friction coefficient can be detected. This is due to the increasing thickness and the contribution to the load-carrying capacity as a result of the pressurized lubricant flow between the reciprocating surfaces. In this regime the tribological properties are very sensitive to both the interaction between the asperities and the operating conditions concerning the lubricant

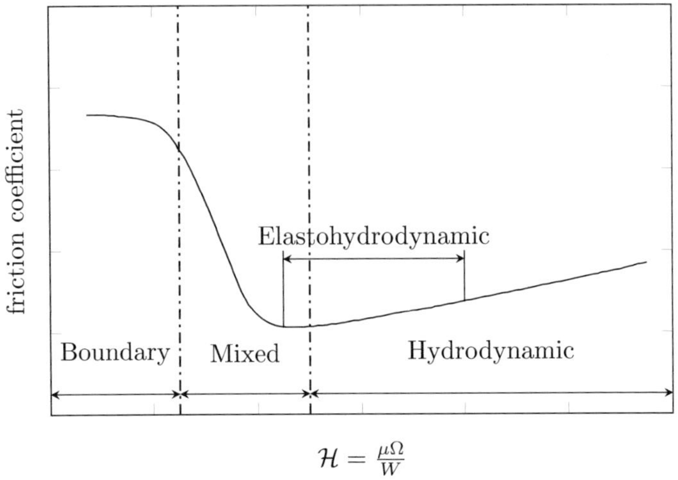

Figure 2.1: Stribeck curve as a function of the Hersey number $\mathcal{H}$. The friction coefficient is computed as the ratio between the tangential and the normal force acting on one of the reciprocating surfaces. Adapted from Hamrock *et al.* [71].

flow. As the Hersey number increases again, the friction coefficient reaches a lower plateau which corresponds to the onset of the hydrodynamic regime. In this regime the two surfaces are adequately separated by the lubricant film and the contact between asperities is negligible. Even if the two surfaces are well separated, elastic deformations can still be induced by pressure distribution in the fluid film. Therefore a further distinction can be made between the hydrodynamic regime and the elastohydrodynamic one [71].

As mentioned before, several studies have proven that surface textures work mainly in the mixed and hydrodynamic lubrication regime [19, 26, 38, 72]. This is because, among the several benefits which are introduced by surface textures, the creation of an additional hydrodynamic lift is the most prominent one and also the one which can better be captured by theoretical models [16]. For this reason, the present work focuses only on the hydrodynamic regime and further analysis in the mixed lubrication region are left to future investigations.

Nonetheless, the modeling of the hydrodynamic regime requires the accurate analysis of the impact of many physical phenomena, such as inertia, thermal effects and cavitation. After presenting the geometry of the tribometer, this chapter focuses on the modeling of the three above-mentioned physical phenomena in the hydrodynamic regime.

2.1 Pin-on-disc tribometer

A tribometer is an instrument which allows to measure tribological quantities, such as friction and wear in a controlled and reproducible manner. It is based on the basic principle of reproducing a sliding contact in a simplified way by focusing on the interaction of two surfaces whose characteristics are representative for more complex designs.

There exist different configurations of tribometers, depending on which of the two surfaces is set in motion and the shape of the indented specimen. The configuration considered in this work is shown in figure 2.2 and consists of a rotating disc whose shaft is loaded in the axial direction by a prescribed normal force F_N. The pin is underneath the disc and is fixed to the lower part of the set-up through a self-aligning support which allows the pin to pitch in order avoid misalignments during the mounting procedure of a new pin. This configuration is the so-called pin-on-disc and is widely used for the study of surface textures, since it allows to change the specimen more easily than other experimental instruments.

According to its definition, the friction coefficient C_f is measured as the ratio between the tangential force F_T resulting from the contact and the force F_N applied on the rotating shaft:

$$C_f = \frac{F_T}{F_N}. \tag{2.2}$$

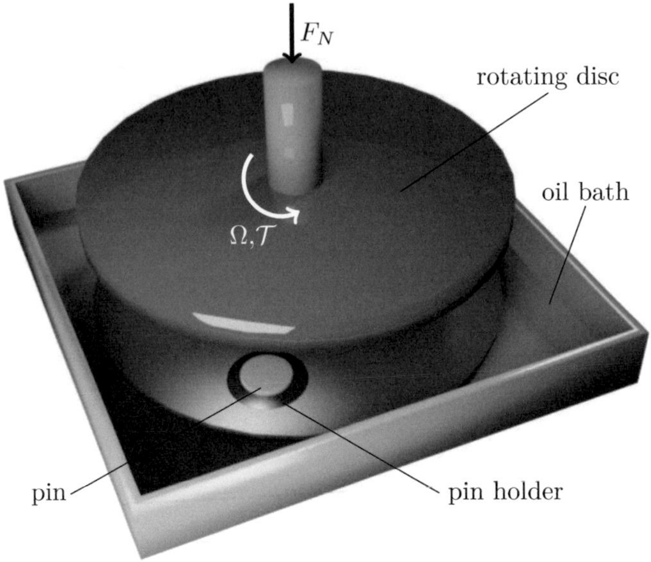

Figure 2.2: Schematic representation of the tribometer used in the experimental investigations ([GMB+15, CFM+18], [19, 20]) and analyzed in the present work.

The tangential force can either be measured through a torque measurement on the shaft or directly through the measurement of the force acting on the pin.

Even though two tribometers are used in the experimental campaigns considered in this work, namely a Plint TE-92 HS (from Phoenix Tribology ltd) and a CSEM (CESM Instruments, Peseux, Switzerland), the principle of the measurement chain is the same. The shaft connected to the disc includes an in-line force transducer for the measurement and feedback control of the load. The tangential force is measured through a strain gauge force transducer mounted on a bracket attached to the upper plate. A temperature sensor provides the feedback for the control of the oil temperature, which can be set up to 200 $^\circ C$.

In both tribometers, new lubricant oil is constantly fed into the contact through a duct in the lower part which is placed in front of the specimen. Although the presence of this feeding line is very important in order to assure the reproducibility of the experimental measurements, its hydrodynamic influence is negligible and will not be considered in the numerical representation of the set-up.

The operating and geometrical parameters that characterize the experimental activity with a tribometer can be classified in three main groups as follows:

- *Tribometer's settings*: the two first operating inputs are the normal force F_N applied on the disc and the angular velocity Ω of the shaft. A third geometrical parameter concerns the distance between the specimen position and the center of rotation of the disc. This distance is also referred to as the sliding radius R_s (see [GMB$^+$15]) and its influence is analyzed in detail in section 2.3.1.
- *Lubricant*: the oil bath can be kept at a prescribed temperature during the running of the experiments. That way, the viscosity μ of the lubricant can be fixed.
- *Specimen/pin*: the most relevant parameter of the pin is the shape of the surface, which can be either nominally flat or slightly curved as shown in Figure 2.3. In case of a curved surface, an important parameter is the shoulder height s_h, which is defined in figure 2.3 as the height between the center of the pin and the edge. In the definition of the surface shape also the following geometrical parameters play a role: the pin radius R_{pin} and the height of the side of the pin h_{side} which is the distance between the edge of the pin and the lower pin support.

In addition to these parameters, a very crucial one is the gap height H, which cannot be fixed *a priori* since it results from the balance between the pressure distribution in the contact and the applied load. Moreover, the pin is mounted on a self-align holder consisting of a half gimbal mechanism which can pitch. This property of the pin holder is exploited at the beginning of each tribological test when the disc is pressed against the pin in order to assure that the pin surface is perfectly aligned with the upper disc surface.

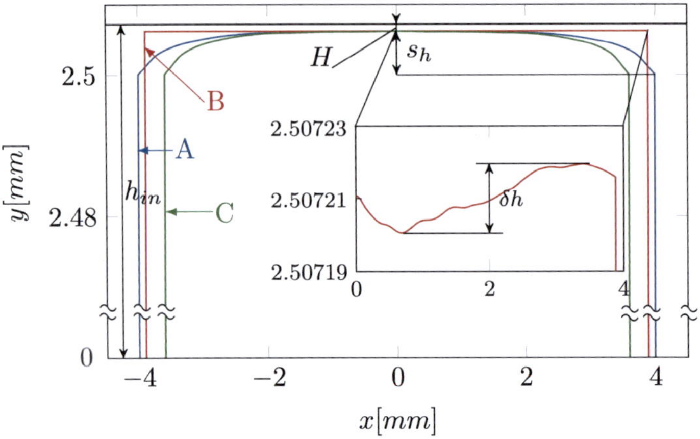

Figure 2.3: Profile of the three different kinds of pin as measured with optic profiler instruments from the experimental set-ups. Pin A is used in the experimental campaigns reported in [19, 20], pin B comes from [CFM$^+$18] and pin C is used in [GMB$^+$15]. Please note that the y-axis is 1000 times magnified. The reference gap height H is defined in the center. The surface profile of pin B is magnified in order to show the roughness profile, the maximal measured height variation is $\delta h = 0.047 \mu m$. The lengths of the zoomed box are expressed in mm.

The alignment is considered to remain stable during the run of the experiments because of the high friction which characterizes the contact between the pin holder and the lower support of the tribometer.

All the textured surfaces considered in this work consist of dimples which are obtained through laser surface texturing (LST). The shape of the dimples is self similar and they resemble the shape of a sphere indented in the surface. The geometrical definition of the texture is explained in detail in chapter 4.

Four experimental campaigns were carried out with the above-described tribometer set-ups. In the following, a brief summary is provided about the main technical aspects of these experimental activities (see also table 2.1), which were subsequently analyzed from the numerical point of view in the present work. The experimental evidences out of the above-mentioned campaigns [19, 20, GMB$^+$15,CFM$^+$18] are then discussed in detail in section 2.3.1 with regard to the preliminary analysis and in chapter 4 focusing on the other experimental campaigns. Three oil lubricants were considered in the experimental campaigns. The numerical modeling of the oil properties is considered in detail in appendix A. The four mentioned analyses are listed in the following:

- **Preliminary analysis**. This preliminary analysis concerns the influence of the sliding radius R_s on the friction performance of both textured and untextured pins. The tribological tests are carried out with two different sliding radii $R_s = 10$ mm and $R_s = 18$ mm. The tribometer from the CSEM instruments was employed, using brass pins and saphire discs, which are proved to have a very low wear rate [44, 73]. An arithmetic mean surface roughness of $R_a = 10$ nm was measured for both the surface of the disc and the pin through atomic force microscopy (AFM). The chosen lubricant oil is the additive-free mineral oil FVA No.1 and the experiments were carried out at room temperature $T = 22°$C. Further details can be found in the work of Greiner *et al.* [GMB$^+$15].
- **Influence of viscosity on the optimal diameter**. In this experimental campaign the Stribeck curve of textured pins with different dimple diameters ranging from 15 μm to 800 μm are investigated at two different temperatures. This analysis is carried out with the Plint TE-92 HS tribometer. The disc consists of steel 100 Cr6 while the pin is made out of normalized steel C85. The lubricant oil employed is an additive-free poly-alpha-olefin PAO-18. The results of this experimental campaign were firstly presented by Braun *et al.* in [19] (2014) and are further explained in chapter 4.
- **Disposition of the texture pattern**. In the recent analysis by Braun et al. [20] (2017), the influence of the dimple disposition is investigated by testing different dimple arrangements ranging from a quadratic disposition to a hexagonal one. This experimental analysis uses the same tribometer settings as the one above. The numerical investigation of this experimental campaign is exposed in section 4.5.

- **Partial texturing**. This last analysis about the impact of partial texturing (see [CFM+18]) employed the CSEM tribometer with a steel 100Cr6 disc. The pin is made out of cemented carbide WC-Ni and was hardened and polished in order to have a very flat surface whose maximal height variation is $\delta h = 0.047 \mu\text{m}$. The lubricant is a automotive oil Shell V-Oil 1404 and the experiments were carried out at room temperature $T = 22°\text{C}$. The corresponding numerical analysis is presented in section 4.3.

2.2 Governing equations of the lubricant film

In the present work, the dynamics of continuum Newtonian fluids in laminar flows is considered. In such a framework, the Navier-Stokes equations represent the most complete way to describe the physics of the fluid, which is mainly characterized by the effects of inertia, pressure and viscosity. There is, however, a category of flows known as "slow viscous motion" in which the pressure and the viscous terms predominate. To this category belong most of the fluid film lubrication problems which are typical of tribological applications in the hydrodynamic regime. For this class of flows the Navier-Stokes equation can be simplified in order to deal with a more compact set of equations, such as the Stokes equations or a scalar equation, such as the Reynolds one. The latter represents one of the most widely used equations in the description of thin film lubrication due to its relative easiness of implementation and the lower computational cost [16].

For what concerns the hypotheses on the fluid and on the flow which are considered in the present work, the following observations have to be made. In the first place, the assumption of continuum fluid is easily fulfilled, as long as the analysis concerns the hydrodynamic lubrication, where, due to the relatively high gap height, the Knudsen number is generally small. Secondly, the hypothesis of laminar flow is also fulfilled, as the maximal Reynolds number (based on the gap height) considered in the present work remains always small enough (see sections 3.2 and 4.4). Finally, the assumption of Newtonian fluid is considered valid for most of the analyses performed, since the shear rate remains relatively small for most of the cases. The validity of this assumption will be further discussed in chapter 4 for what concerns certain

	preliminary analysis	optimal diameter	dimple disposition	partial texture
tribometer	CSEM	Plint TE-92 HS	Plint TE-92 HS	CSEM
disc material	saphire	steel 100Cr6	steel 100Cr6	steel 100Cr6
disc diameter	50 mm	70 mm	70 mm	70 mm
sliding radius	varying	30 mm	30 mm	30 mm
pin type	C	A	A	B
pin diameter	7.2 mm	8 mm	8 mm	7.77 mm
pin material	brass	normalized steel C85	normalized steel C85	cemented carbide WC-Ni
lubricant	mineral oil FVA	PAO-18	PAO-18	Shell V-Oil 1404
normal force F_N	12 N	150 N	150 N	15 N
reference publication	[GMB+15]	[19]	[20]	[CFM+18]

Table 2.1: Survey of the tribometer set-ups out of the experimental campaigns whose numerical representation is considered in this work.

particular conditions which are taken into account in the parametric study about surface textures.

In the following, the governing equations for the modeling of the fluid film lubrication are presented in a top-down approach, starting from the most complete one, namely the compressible Navier-Stokes equations, to the most compact and computationally efficient one, namely the Reynolds equation.

2.2.1 Navier-Stokes equations

Both the compressible and the incompressible Navier-Stokes equations are considered in the present work for a steady flow. The steady-state compressible Navier-Stokes equations are a partial differential set of equations which represent the conservation of mass, momentum and energy and can be written as follows:

$$\begin{cases} \nabla \cdot (\rho \mathbf{u}) = 0 & \text{(2.3a)} \\ \rho (\mathbf{u} \cdot \nabla) \mathbf{u} + \nabla p = \nabla \cdot \sigma & \text{(2.3b)} \\ \nabla \cdot \left[\left(\rho e^t + p \right) \cdot \mathbf{u} \right] = \nabla \cdot (k_T \nabla T + \mathbf{u} \cdot \sigma) & \text{(2.3c)} \end{cases}$$

where p,T and ρ are pressure, temperature and density, respectively while $\mathbf{u}$ is the three dimensional velocity vector. The variable e^t is the specific total energy and can be expressed as $e^t = C_v T + \frac{1}{2} |\mathbf{u}|^2$ for a fluid whose specific heat capacity C_v does not depend on temperature. The viscous stress tensor can be expressed as $\sigma = \left(2\mu \mathcal{D}(\mathbf{u}) + \hat{\lambda}(\nabla \cdot \mathbf{u})\right)$, where μ is the dynamic viscosity and $\hat{\lambda} = -\frac{2}{3}\mu$ is the dilatational viscous coefficient and k_T is the temperature diffusion coefficient.

In order to be closed, the set of equations 2.3 also needs to be coupled with an equation of state (EOS) for the fluid $\rho(p,T)$ and a constitutive law for the dependence of the viscosity on pressure and temperature $\mu(p,T)$. Two equations of state are considered in this work, namely the stiffened gas equation of state and the Dowson-Higginson one while the lubricant is assumed to behave as a Newtonian fluid whose viscosity varies exponentially with temperature. Both equations of state and the viscosity constitutive law are presented in detail in appendix A for the three different lubricant oils which are considered in this work.

The set of equations 2.3 is discretized through the finite volume method and implemented in the in-house solver KIT-SPARC [74]. The code is implemented in FORTRAN 2003 while the parallelization relies on the message passing interface (MPI) and is based on a multi-block structured grid. In order to converge faster to the steady state, the solver employs a full multigrid approach and the pseudo time integration is done using a 4th order explicit Runge-Kutta method. Steady state simulations are carried out until the residuum drops below 10^{-5} on the finest grid. Both convective and diffusive fluxes are discretized in space with a second order accuracy scheme.

The use of a compressible approach for such slow motion problems can lead to stiffness-related problems due to the hyperbolic nature of the compressible Navier-Stokes equations. In order to cope with this well-known issue [75], a preconditioner is employed with the aim of obtaining a stable and fast convergence even at the very low Mach numbers which characterize typical tribological systems.

Particular attention has to be paid in the assignment of the boundary conditions of the compressible Navier-Stokes equations, since they cannot be freely prescribed [76]. In the simulations considered in this work, the flow remains always subsonic. Therefore only four out of five boundary conditions can be prescribed at the domain inlet while the fifth one has to be assigned at the outlet. Further details about the mesh generation and the boundary conditions are given case by case while presenting the simulations carried out with the compressible Navier-Stokes equation in sections 2.3.1, 2.3.2 and 4.1.

Also the incompressible Navier-Stokes equations are considered in the present work, mainly with the aim to study the impact of inertia in the texture scales. The incompressible set of equations can be easily obtained from the set of equations 2.3 under the hypothesis of a constant density $\rho = \bar{\rho}$. Moreover, no temperature variations are considered in order to focus better on the effect of the inertial terms only. Under this assumptions we can rewrite the equations 2.3 as follows:

$$\begin{cases} \nabla \cdot \mathbf{u} = 0 & \text{(2.4a)} \\ \rho \left(\mathbf{u} \cdot \nabla \right) \mathbf{u} + \nabla p = \mu \nabla^2 \mathbf{u}. & \text{(2.4b)} \end{cases}$$

The incompressible Navier-Stokes equations are solved numerically with the commercial software ANSYS FLUENT® [77]. The solver is based on the finite volume method and employs the SIMPLE pressure-velocity coupling algorithm.

2.2.2 Reynolds equation

Fluid dynamic computational models, which consider the three dimensional geometry of lubrication problems and also the nonlinear effects induced by inertia, are often computational demanding due to the complexity of the solution and the high amount of degree of freedom. For this reason it is always preferable to simplify the modeling of the fluid flow whenever possible. A widely used simplification in the field of tribology concerns the determination of which length scales play a more relevant role in order to find out which terms of the Navier-Stokes equation can be dropped out. In this regard, the typical applications in tribology consist of a thin lubricant film squeezed between two sliding surfaces whose characteristic lengths in the streamwise and spanwise directions (L_x and L_z respectively) are much bigger than the gap height h in the wall normal direction y (see figure 2.4). These conditions can be generalized and taken as hypothesis in order to simply the Navier-Stokes equation.

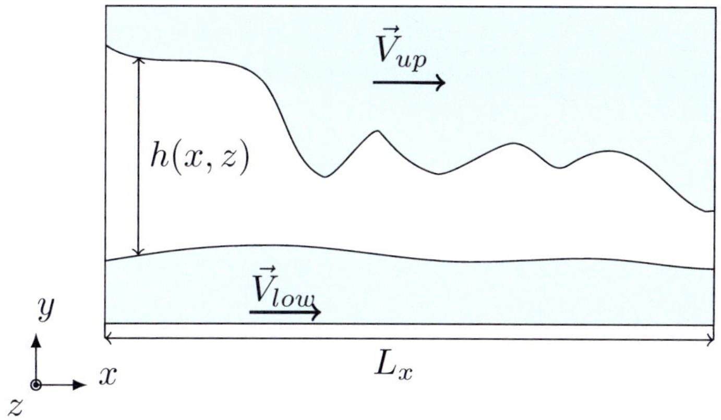

Figure 2.4: Schematic representation of a generic gap height distribution $h(x,z)$ with the reference system and lengths.

Moreover, a second assumption can usually be considered, by assuming that the two surfaces have a certain regularity and the slopes $\frac{\partial h}{\partial x}$ and $\frac{\partial h}{\partial z}$ are small. Under these two hypotheses, the steady Navier-Stokes equations 2.4 can be simplified to the Reynolds equation as follows:

$$\nabla \cdot \left(h^3 \nabla p - 6\mu \vec{V} h \right) = 0 \tag{2.5}$$

where $\vec{V}$ is defined as the relative velocity vector between the velocity of the upper and lower surface $\vec{V} = \vec{V}_{up} - \vec{V}_{low}$. The Reynolds equation is a scalar partial differential equation whose unknown is the pressure field $p(x,z)$.

As pointed out by Dowson [78] and also other authors [79], the two above-mentioned hypotheses about the scales and regularity of the two surfaces, are necessary and sufficient to derive the Reynolds equation. Nonetheless, there is another important condition which has to be verified in order to assure the applicability of the Reynolds equation also in practical applications. This condition concerns the Reynolds number, which is generally defined as follows, based on the reference gap height H:

$$Re = \frac{\rho H V}{\mu} \tag{2.6}$$

where ρ is the density of the fluid, V the magnitude of the relative velocity and μ the dynamic viscosity. As a matter of fact, a limitation in the applicability of the Reynolds exists when the Reynolds number exceeds a certain threshold, because of the increasing relevance of inertial effects and the possible occurrence of turbulence. This limitation, as well as the one about the regularity of the surfaces, were analyzed in detail by Dobrica and Fillon in [46] with the 1D Reynolds equation. Section 2.3.3 presents an extension of Dobrica and Fillon's work for the 2D Reynolds equation applied to the texture geometry analyzed in this thesis.

The Reynolds equation presented in the above formulation 2.5 can also be extended in order to model thermodynamic effects and cavitation. In this thesis, these two aspects of the Reynolds equation are modeled separately as shown in the next paragraphs.

It is important to notice that all the models based on the Reynolds equation presented in this work are discretized through the finite volume method. As underlined by Wolozynski *et al.* in [80], the use of the finite volume method represents noticeable advantages in terms of stability and convergence speed in comparison to the finite difference method. This property of the finite volume method becomes particularly relevant in the presence of discontinuities in the gap height distribution, for example at the edges of the pin (see figure 2.3) or in case of textured surfaces. This is due to the fact that the finite volume method discretizes the so-called weak formulation of the Reynolds equation as shown in equation 2.5. Therefore, this approach requires to discretize only the pressure gradient while the finite differences method involves the discretization of the pressure Laplacian and the gradient of the gap height. On the other side, the finite volume method requires a slightly more complex grid, a cell-based mesh is needed in which the pressure is computed at the node in the center of the cells while the fluxes have to be computed at the cell interfaces. The finite volume discretization used in this work is based on the approach shown by Arghir *et al.* in [81].

The Reynolds equation 2.5 can be expressed also through its non-dimensional formulation by introducing the reference variables μ_r, h_r, p_r, and U_r:

$$\tilde{\nabla} \cdot \left(\tilde{h}^3 \tilde{\nabla} \tilde{p} - 6 \frac{\mu_r U_r L_x}{h_r^2 p_r} \tilde{\mu} \tilde{\vec{V}} \tilde{h} \right) = 0 \tag{2.7}$$

where the letters marked with $(\tilde{\cdot})$ express the non-dimensional variables. The parameters group before the second term is referred to as the Sommerfeld number $\mathcal{S}$ (or also bearing number) [71]. This number can also be rewritten, without loss of generality, by using the reference applied load W_r instead of the reference pressure p_r:

$$\mathcal{S} = \frac{\mu_r U_r L_x}{h_r^2 W_r}. \tag{2.8}$$

The Sommerfeld number is widely used to analyze the operating conditions of tribological applications and it will extensively be used to generalize the effects of viscous dissipation and surface textures in chapter 3 and 4, respectively. Thanks to the formulation of the Sommerfeld number, one can rewrite equation 2.9 in following way:

$$\tilde{\nabla} \cdot \left(\tilde{h}^3 \tilde{\nabla} \tilde{p} - 6 \mathcal{S} \tilde{\mu} \tilde{\vec{V}} \tilde{h} \right) = 0. \tag{2.9}$$

Thermal and compressible formulation

For the purpose of taking into account the variation of the thermodynamic quantities, such as pressure, density and temperature, the Reynolds equation

2.5 can be coupled with the energy equation and a suitable equation of state which describes the thermodynamic properties of the lubricant. The 1D formulation of this coupling is considered in this work with the main purpose of studying the impact of viscous dissipation on the pressure losses and on the temperature increase in thin film lubrication (see chapter 3). The current formulation of the coupling can be derived from the compressible Navier-Stokes equations 2.3 under the following hypotheses:

1. the film thickness h is much smaller than the characteristic length of the two surfaces, namely L_x and L_z;
2. the nonlinear terms are negligible in comparison to viscous ones (low Reynolds number);
3. the viscosity μ and thermodynamic variables p, ρ and T are constant in the wall normal direction y.

Variations of viscosity and the thermodynamic variables along the channel length x are considered. Starting from the above-mentioned hypotheses, one can write the 1D Reynolds equation in its steady and compressible formulation as follows:

$$\frac{d}{dx}\left[\frac{\rho h^3}{\mu}\frac{dp}{dx}\right] - 6U\frac{d}{dx}(\rho h) = 0. \tag{2.10}$$

In many studies concerning the coupling between the Reynolds equation and the energy one, the latter is discretized also in the wall normal direction [48, 67, 82, 83]. This typically leads to a big increase in the computational cost, because of the additional dimension in which the energy equation has to be discretized. In order to obtain a form of the energy equation which is more suitable for coupling with the Reynolds equation, one can perform an integration in the wall normal direction, provided that the above-mentioned hypotheses are fulfilled. This procedure is similar to the one which leads to the derivation of the Reynolds equation.

By recalling the energy equation 2.3c one can introduce the total enthalpy H^t as follows:

$$H^t = \rho e^t + p = \rho e + \frac{1}{2}\rho|\mathbf{u}|^2 + p \tag{2.11}$$

where e is the specific internal energy, which linearly depends on temperature according to the relation $e = C_v T$. Moreover, in tribological flows the kinetic energy contribution can be neglectable with good approximation. Therefore, we can restrict our analysis to the internal enthalpy only:

$$H(T,p) = \rho C_v T + p. \tag{2.12}$$

The energy equation 2.3c can now be rewritten in its 1D formulation as function of the internal enthalpy.

$$\frac{d}{dx}[Hu] = k_T \frac{d^2T}{dx^2} + \mu\left(\frac{\partial u}{\partial y}\right)^2 \tag{2.13}$$

where u is the streamwise component of the velocity $\mathbf{u}$.

The next step consists in substituting the Couette-Poiseuille velocity profile obtained by the previous considerations about the continuity and the momentum equation,

$$u(x,y) = \frac{1}{2\mu}\frac{\partial p}{\partial x}\left(y^2 - yh\right) + \frac{U}{h}y \tag{2.14}$$

into the energy equation

$$\frac{d}{dx}\left[H\left(\frac{1}{2\mu}\frac{\partial p}{\partial x}\left(y^2 - yh\right) + \frac{U}{h}y\right)\right] = k_T\frac{d^2T}{dx^2} + \mu\left(\frac{1}{2\mu}\frac{\partial p}{\partial x}(2y-h) + \frac{U}{h}\right)^2. \tag{2.15}$$

Where U denotes the velocity of the upper wall in the x-direction. The above expression can now be integrated in the wall normal direction y thanks to the hypotheses of thin film lubrication. Finally, the integrated energy equation reads

$$\frac{d}{dx}\left[H\left(-\frac{h^3}{12\mu}\frac{dp}{dx}+\frac{Uh}{2}\right)\right]=kh\frac{d^2T}{dx^2}+\frac{h^3}{12\mu}\left(\frac{dp}{dx}\right)+\mu\frac{U^2}{h}. \tag{2.16}$$

Similar derivations of the energy equation for thin films can also be found in other works [84–86]. Nonetheless, the current formulation is rather versatile and general, since only the kinetic energy term $\frac{1}{2}\left|\mathbf{u}^2\right|$ is neglected as additional hypothesis to those done for the Reynolds equation. Moreover, in comparison to the cited similar works, the terms involving the pressure transport $\frac{d}{dx}(p\mathbf{u})$ and the viscosity gradients in the streamwise direction remain in the equation, hence addressing the problem with a more general approach.

Particular attention deserve the last two terms of equation 2.16 which represent the viscous dissipation and, hence, the mechanical power which is lost from the motion of the fluid and converted into thermal power. In other words, these two terms correspond to the main coupling mechanism between the Reynolds equation and the energy equation. The practical implication due to the influence of these terms will be further discussed in section 3.3.

Since the internal enthalpy directly depends on temperature and pressure, one can solve the coupling between the Reynolds and the energy equation through an iterative process, whose details are shown in figure 2.5. In this way, each equation can be discretized with respect to a single thermodynamic unknown, *i.e.* the pressure for the Reynolds equation and the temperature for the energy equation. Therefore, the following two linear systems for pressure and temperature are obtained respectively:

$$A_p p = b_p \tag{2.17}$$

and

$$A_T T = b_T. \tag{2.18}$$

Both of the above linear systems are obtained through a discretization based on the finite volume method.

The convergence of the iterative loop is controlled on the normalized residues of both pressure ϵ_p and temperature ϵ_T fields. The normalized residue at the iteration step n is computed through the following definition:

$$\epsilon_p = \frac{\sum_{i=1}^{N_x} |p_i^n - p_i^{n-1}|}{\sum_{i=1}^{N_x} |p_i^{n-1}|}. \tag{2.19}$$

The validity of this formulation of the compressible and thermal Reynolds equation is further discussed in chapter 3 by comparing it to the solution of the compressible Navier-Stokes equation. In particular, two non-dimensional numbers are considered in order to evaluate the applicability conditions of the proposed hypotheses, namely the Reynolds number (see Eq. 2.6) and the Péclet number. The latter expresses the ratio of the convected heat by the fluid motion to the heat diffusion through the fluid itself. This quantity can be derived from the non-dimensionalization of the energy equation 2.3c and is related to the temperature distribution in thin film lubrication [66]

$$Pe = \frac{\rho C_p U h}{k_T} \tag{2.20}$$

where C_p is the specific heat coefficient at constant pressure.

Cavitation model

The cavitation model used in the present work is the so-called Fischer–Burmeister–Newton–Schur (FBNS) presented by Woloszynski *et al.* in [69]. The starting point is the Reynolds equation formulated in terms of pressure p and cavity fraction θ as proposed by Elrod and Adams in [87] which is, in turn,

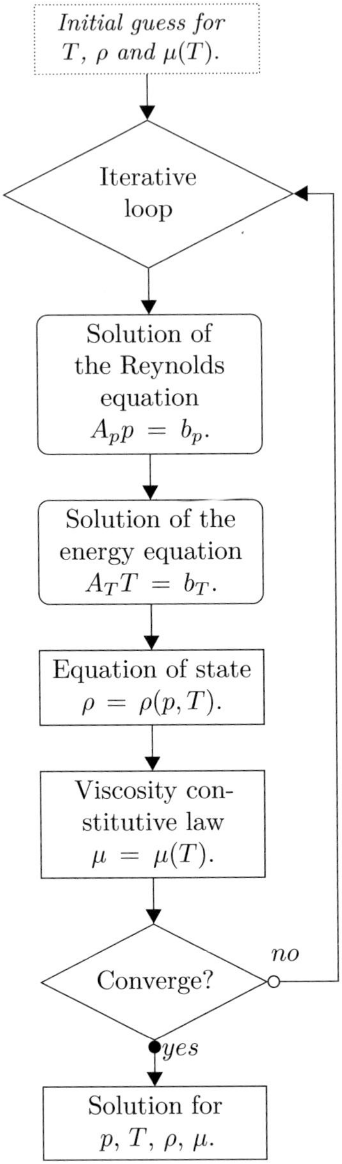

Figure 2.5: Flowchart of the iterative solution of the thermal compressible Reynolds equation.

based on the cavitation model presented by Jakobsson, Floberg and Olsson (JFO) [88, 89]. The Reynolds equation including the cavity fraction reads:

$$\nabla \cdot \left(h^3 \nabla p - 6\mu \vec{V} h (1-\theta) \right) = 0. \tag{2.21}$$

The cavity fraction θ is defined through a reference density ρ_{ref} as follows:

$$\theta = 1 - \frac{\rho}{\rho_{ref}}. \tag{2.22}$$

The so-defined cavity fraction is zero when the flow is completely in the liquid phase and its maximum value is one when the fluid is completely in the vapor phase. At the same time, the difference between the pressure and the cavitation pressure p_{cav} in the cavitation zone is zero. These two aspects can be combined into a single condition which constitutes the complementarity constraint of equation 2.21:

$$(p - p_{cav})\,\theta = 0 \tag{2.23}$$

where both the relative pressure $p - p_{cav}$ and the cavity fraction θ are positive quantities.

The cavitation problem is usually solved iteratively by solving the Reynolds equation 2.21 and then correcting the results through the application of the constrain 2.23. Even though this solving approach is the most pursued in literature [68, 90, 91], [Lop17], the newly proposed FBNS algorithm presents noticeable advantages in terms of convergence speed and stability in the presence of gap height discontinuities [69]. The reason lies in the fact that in the FBNS algorithm both the Reynolds equation 2.21 and the constraint 2.23 are rewritten in a single one leading to an unconstrained optimization problem.

In order to derive such a formulation the constraint equation 2.23 can be reformulated using the so-called Fischer–Burmeister equation:

$$F_i = F_i(p_i, \theta_i) = p_i + \theta_i - \sqrt{p_i^2 + \theta_i^2} = 0 \qquad \text{for} i = 1, ..., N \tag{2.24}$$

where N is the number of computational nodes in the domain. In this way, the constraint condition expressed in equation 2.23 is replaced by a system $\mathbf{F} = \mathbf{F}(\mathbf{p}, \theta) = 0$ of N equations, where $\mathbf{p}$ and θ are the discretized pressure and cavity fraction vectors. The Reynolds equation 2.21 can also be rewritten in a similar way as an implicit relation $\mathbf{G} = \mathbf{G}(\mathbf{p}, \theta) = 0$. The resulting reformulation consists in an unconstrained system of $2N$ equations out of the coupling between $\mathbf{F} = \mathbf{F}(\mathbf{p}, \theta) = 0$ and $\mathbf{G} = \mathbf{G}(\mathbf{p}, \theta) = 0$.

An efficient solution of the unconstrained system can be obtained by means of the Newton–Raphson method. The solution is computed iteratively by considering the incremental update expressed by $\mathbf{p}^{(k+1)} = \mathbf{p}^{(k)} + \delta \mathbf{p}^{(k)}$ and $\theta^{(k+1)} = \theta^{(k)} + \delta \theta^{(k)}$, where $\delta \mathbf{p}^{(k)}$ and $\delta \theta^{(k)}$ are the solution updates at the k-th iteration. The solution updates are computed by considering the following system based of the Jacobian matrices of $\mathbf{F}$ and $\mathbf{G}$:

$$\mathbf{J} \begin{pmatrix} \delta \mathbf{p}^{(k)} \\ \delta \theta^{(k)} \end{pmatrix} = \begin{bmatrix} \mathbf{J}_{\mathbf{F},\mathbf{p}} & \mathbf{J}_{\mathbf{F},\theta} \\ \mathbf{J}_{\mathbf{G},\mathbf{p}} & \mathbf{J}_{\mathbf{G},\theta} \end{bmatrix} \begin{pmatrix} \delta \mathbf{p}^{(k)} \\ \delta \theta^{(k)} \end{pmatrix} = - \begin{pmatrix} \mathbf{F}^{(k)} \\ \mathbf{G}^{(k)} \end{pmatrix}. \tag{2.25}$$

The solution of this system now presents two main challenges, namely that the matrix $\mathbf{J}$ is typically badly conditioned and that the system size is twice the original one. The first mathematical issue can be overcome by preconditioning the matrix $\mathbf{J}$. This operation can easily be done by comparing the elements in the matrices $\mathbf{J}_{\mathbf{F},\mathbf{p}}$ and $\mathbf{J}_{\mathbf{F},\theta}$ and swapping the respective columns whenever

$J^i_{\mathbf{F},\mathbf{p}} < J^i_{\mathbf{F},\theta}$. In this way, the matrix $\mathbf{J}$ is reordered into the new matrix $\mathbf{J}_r$ and equation 2.25 can be rewritten as:

$$\mathbf{J}_r \begin{pmatrix} \delta\mathbf{a} \\ \delta\mathbf{b} \end{pmatrix} = \begin{bmatrix} \mathbf{A_F} & \mathbf{B_F} \\ \mathbf{A_G} & \mathbf{B_G} \end{bmatrix} \begin{pmatrix} \delta\mathbf{a} \\ \delta\mathbf{b} \end{pmatrix} = - \begin{pmatrix} \mathbf{F} \\ \mathbf{G} \end{pmatrix} \tag{2.26}$$

where $\delta\mathbf{a}$ and $\delta\mathbf{b}$ are the reordered solution updates and $\mathbf{A_F}$, $\mathbf{B_F}$, $\mathbf{A_G}$ and $\mathbf{B_G}$ are the preconditioned Jacobian matrices.

In order to overcome the issue related to the extended size of the Jacobian matrix $\mathbf{J}_r$, the system is solved by substitution in two steps, hence by splitting the problem as follows:

$$\begin{aligned} \left(\mathbf{B_G} - \mathbf{A_G}\mathbf{A_F}^{-1}\mathbf{B_F}\right)\delta\mathbf{b} &= -\mathbf{G} + \mathbf{A_G}\mathbf{A_F}^{-1}\mathbf{F} \\ \mathbf{A_F}\delta\mathbf{a} &= -\mathbf{F} - \mathbf{B_F}\delta\mathbf{b}. \end{aligned} \tag{2.27}$$

The matrix $\mathbf{B_G} - \mathbf{A_G}\mathbf{A_F}^{-1}\mathbf{B_F}$ is typically non-singular and ensures a superlinear convergence of the Newton method if the initial condition is close enough to the solution [69, 92]. This can easily be achieved at least for the pressure initial guess by solving at first the Reynolds equation without cavitation (Eq. 2.5).

After solving the system in equation 2.27 for $\delta\mathbf{a}$ and $\delta\mathbf{b}$, one can reconstruct the original updates vectors $\delta\mathbf{p}$ and $\delta\theta$ through a reverse swapping and finally obtain the updated solution for $\mathbf{p}$ and θ. The new solution can now be used to compute the new Jacobian matrices $\mathbf{J}_{\mathbf{F},\mathbf{p}}$ and $\mathbf{J}_{\mathbf{F},\theta}$ for the next iteration of the Newton algorithm. The iteration process is stopped when the norm of the update vectors $\delta\mathbf{p}$ and $\delta\theta$ get below a certain tolerance. In the present work the tolerance is set to 10^{-6} which typically requires between 10 and 15 iterations of the FBNS algorithm.

The Reynolds equation expressed in the form $\mathbf{G}(\mathbf{p},\theta) = 0$ is numerically discretized in a similar way as shown before for the classic Reynolds equation (section 2.2.2). The only difference concerns the fact that the equation becomes hyperbolic in the cavitated area when a mass-conserving algorithm is included. For this reason, the discretization requires a first order upwind scheme for the Couette term of equation 2.21. Even though this may represent a limitation for the convergence of the spatial scheme, the high converging speed of the overall scheme still makes this algorithm competitive in comparison to other more used ones [69]. As a matter of fact, the solution of the FBNS algorithm with a mesh of 2048x2048 points requires only a few minutes, making this algorithm very suitable for extensive parametric studies, such as those presented in chapter 4.

2.3 Preliminary analysis in the hydrodynamic regime

This section presents the preliminary analysis performed on the tribometer's geometry. The goal is to investigate the effects of geometrical and physical aspects of the experimental set-up in order to identify the most suitable numerical framework for a performing analysis of surface texture. To achieve this goal, this analysis focuses on two scales, namely a large one regarding the macroscopic geometry of the experimental set-up and a second one concerning the micro scale of a typical texture element. The investigation in the macro scales has the double aim to describe the effects of the sliding radius R_s (*i.e.* of the velocity gradient in the rotation plane) and to assess the role of inertial effects in the analysis of the untextured pin. Subsequently, the analysis on the single texture element concerns the comparison between the Navier-Stokes equations and the Reynolds one in order to verify the applicability conditions of the latter expressed in section 2.2.2.

2.3.1 Velocity gradient effects

The sliding radius R_s represents an important operating parameter in the carrying out of experiments with a pin-on-disc set-up. In most of the cases, an

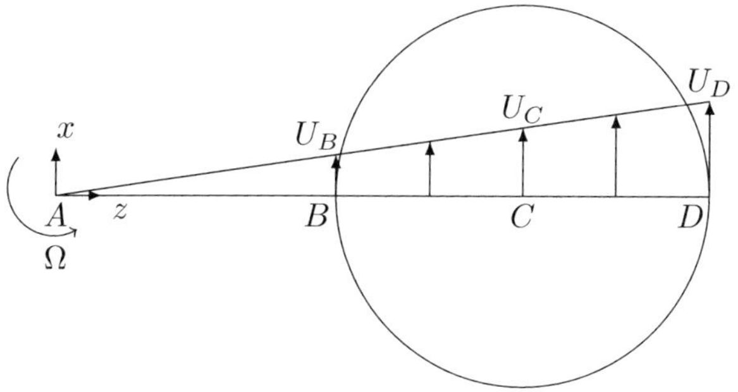

Figure 2.6: Schematic representation of the velocity profile in the rotation plan x-z. The sliding radius is defined as the distance between the center of the pin and the center of rotation of the disc $R_s = z_C - z_A$.

ideal uniform distribution of velocity in the rotation plane x-z is preferred, in order to reproduce the tribological condition in a more homogeneous and reproducible way. Nonetheless, due to the fact that the sliding conditions are obtained through the rotation of the disc, the velocity distribution over the pin is always a function of distance form the center of rotation (*i.e.* the sliding radius R_s). This leads to a difference in the sliding speeds over the pin surface which is referred to as velocity gradient. Even if mostly unwanted, velocity gradient effects are experienced by many applications, for example unidirectional contacts, such as typical thrust bearings or start-stop systems in modern automobiles [93, 94].

The impact of the velocity gradient on the tribological performance of both textured and untextured pins was investigated experimentally and numerically in the work by Greiner *et al.* [GMB+15]. This section presents more in detail the numerical analysis carried out by the author of the present thesis for above-mentioned paper.

Figure 2.6 shows the schematic representation of the velocity profile induced by the rotation of the disc over the pin. Based on the work of Greiner *et al.* [GMB+15], the velocity gradient is defined as the velocity difference between the inner and the outer side of the pin normalized by the pin radius

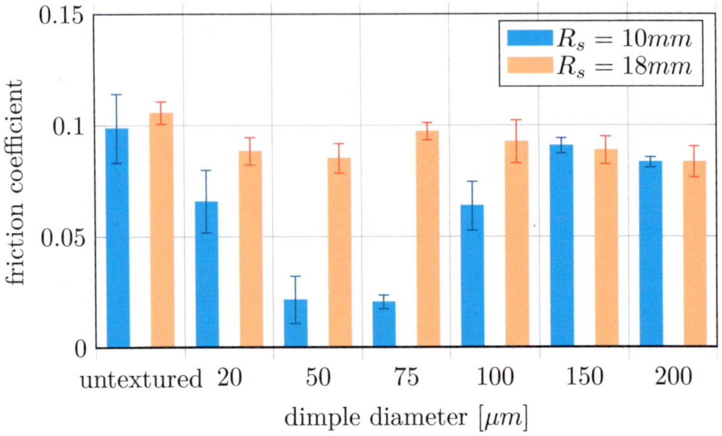

Figure 2.7: Experimental measurements of the friction coefficient as a function of dimples diameter and distance from the center of rotation [GMB+15]. The error bars represent the standard deviation. The untextured case is compared to different diameters ranging from 20 to 200μm while the dimple depth is kept constant to $h_D = 20\mu m$.

$\Delta U = \frac{U_D - U_B}{R_{pin}}$. The sliding radius is defined as the distance between the center of the pin and the center of rotation $R_s = z_C - z_A$. In the experiments carried out in [GMB+15], two different sliding radii are investigated, namely $R_s = 10$ mm and $R_s = 18$ mm. In both cases the centerline velocity U_C, which corresponds to the average velocity, is kept constant by changing the angular velocity of the disc Ω, so that in both cases $U_C = 0.1$ m/s. That way, the velocity gradient ΔU is smaller in case of a larger sliding radius. The textured surfaces were composed by equally spaced spherical dimple in a hexagonal pattern. Different dimple diameter, ranging from 20 to 200 μm, were tested and compared to the untextured case, while the dimple depth was kept constant to $h_D = 20$ μm and the texture density was also kept constant to $\rho_{txt} = 22\%$. Further details about the experiments are reported in table 2.1.

The main results of the experimental activity carried out in [GMB+15] are shown in figure 2.7. The figure shows the measured friction coefficient as a function of the dimple diameter (as well as the reference untextured case) for

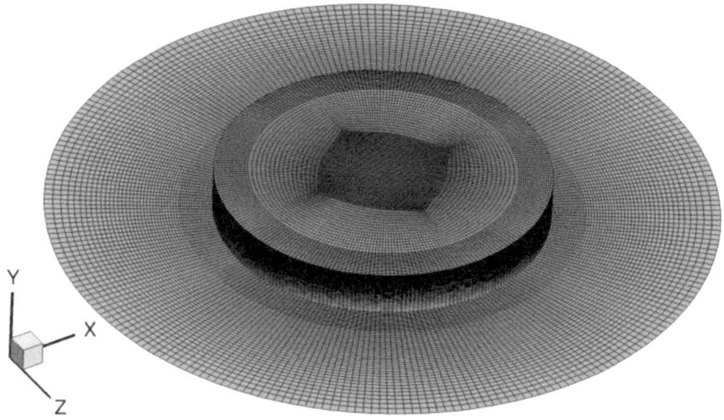

Figure 2.8: Representation of the multigrid block-structured mesh used for the simulations with the compressible Navier-Stokes equations. In order to ease the graphic representation, the shown mesh corresponds to the fourth multigrid level out of five, whereas the number of cells is doubled in each direction when the multigrid level is increased. Moreover, the upper rotating wall is not shown and the y-axis is magnified 10 times. The fifth multigrid level consists of about $1.2 \cdot 10^6$ cells

both of the investigated sliding radii R_s. All the tests carried out with the larger sliding radius R_s show a higher friction coefficient, also in the untextured case. Moreover, the effects of the velocity gradient lead to different values of dimple diameters which corresponds to the minimum friction coefficient.

Parallel to the experimental activity, numerical simulations of the untextured pin for both sliding radii were carried out. The numerical analysis consider the real pin geometry as measured through an optical profiler scan (see figure 2.3). The simulations are based on the compressible Navier-Stokes equations as shown in section 2.2.1 which are solved with the in-house code KIT-SPARC. Figure 2.8 shows the multigrid block-structured mesh used in the simulations. The numerical analysis is restricted to the hydrodynamic lubrication regime. For this reason, the minimum gap height considered in this preliminary study is not lower than $H = 1\mu m$ and it is anyway larger than the measured arithmetic roughness $R_a = 10$ nm in order to neglect the effects of roughness.

As boundary condition, the lower part of the domain and the whole pin surface are considered as static wall with no-slip. The upper wall has a prescribed velocity $\mathbf{u}(x,z,y=\bar{y}) = \Omega \times \mathbf{r}$, where $\bar{y}$ is the y coordinate of the upper wall and $\mathbf{r}$ is the distance with respect to the center of rotation. The magnitude of the angular velocity is set so that the linear velocity at the center of the pin corresponds to the same one considered in the experiments *i.e.* $U_C = 0.1$ m/s. At the inlet, the total pressure is prescribed equal to the reference ambient pressure $p^t_{in} = 10^5$ Pa since the very small disc velocity makes the dynamic pressure contribution negligible. A static temperature $T_{in} = 22°C$ is prescribed at the inlet in agreement with the room temperature at which the experiments were carried out. Since this numerical analysis falls under the preliminary work of this thesis, cavitation is not directly modeled in the Navier-Stokes equation but is considered through the application of the so-called half Sommerfeld assumption. Therefore the minimum pressure cannot drop below the cavitation pressure, which is set to $p_{cav} = 80000$ Pa.

The resulting pressure field is shown in figures 2.9 and 2.10 for the case of $R_s = 10$ mm and $R_s = 18$ mm, respectively. As one can see, the pressure distribution is influenced by the different sliding radii R_s applied through the boundary condition on the upper wall. In the case of a smaller sliding radius $R_s = 10$ mm, the pressure peak in the first part of the pin is slightly higher than in the second case. This results in a higher normal force (*i.e.* load carrying capacity) which is computed through the integration of the pressure distribution as follows:

$$F_N = \int_A p(x,z)dxdz. \tag{2.28}$$

The computed normal force with $R_s = 10$ mm is $F_N = 11.375$ N while for $R_s = 18$ mm is $F_N = 11.034$ N. Therefore, in the case of a smaller sliding radius the load-carrying capacity is about 3% higher. The friction coefficient

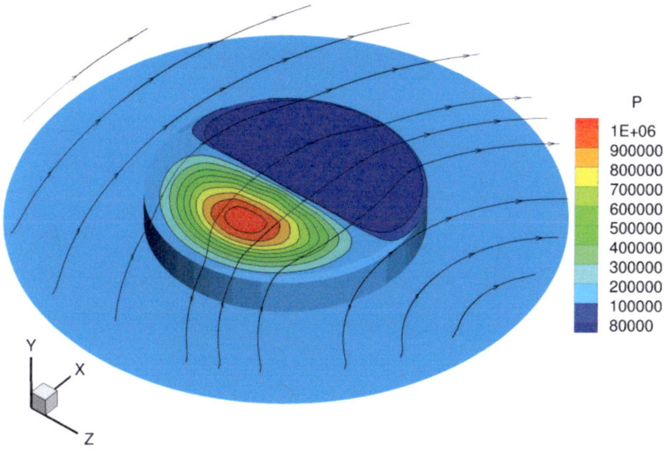

Figure 2.9: Pressure distribution and streamlines obtained with $R_s = 10$ mm. The upper wall is not shown and the y-axis is magnified 10 times.

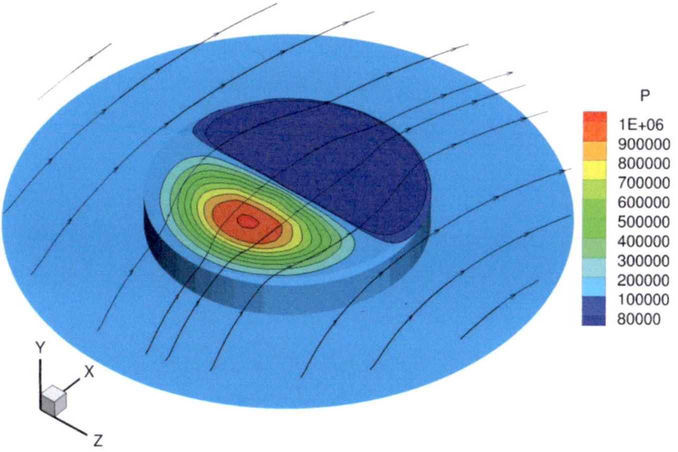

Figure 2.10: Pressure distribution and streamlines obtained with $R_s = 18$ mm. The upper wall is not shown and the y-axis is magnified 10 times.

can be computed through the definition from equation 2.2 which requires, in turn, the computation of the tangential force:

$$F_T = \int_A \tau(z, x, y = \bar{y}) dx dz. \tag{2.29}$$

The computed tangential forces for the case with $R_s = 10$ mm is $F_T = 0.0375$ N which is slightly higher than the one for the case with $R_s = 18$ mm which results in $F_T = 0.0370$ N. Nonetheless, the friction coefficient results smaller for the case with smaller sliding radius $C_{f_{R_s=10}} = 3.301 \cdot 10^{-3}$ than in the case with larger sliding radius $C_{f_{R_s=18}} = 3.351 \cdot 10^{-3}$ because of the much higher normal force. This fact is in agreement with the experimental results shown in figure 2.7 for the untextured case. Moreover, the results presented in this section and in the work of Greiner *et al.* are in compliance with a similar analysis presented by Zhang *et al.* in [37] in which the authors study the influence of the sliding radius with a pin-on-disc tribometer but with a textured disc. As a matter of fact, the sliding radius is a very sensitive parameter whose influence has to be considered while concealing the setting of tribological analyses based on a pin-on-disc set-up.

2.3.2 Inertial effects in the pin-on-disc scale

This section focuses on the Direct Numerical Simulation (DNS) with the compressible Navier-Stokes equations 2.3 which were carried out in order to estimate the relevance of inertia in the hydrodynamic flow of the tribometer set-up. This part of the preliminary analysis has the main goal to quantify the role of the single terms of the Navier-Stokes equations. Similar balance analyses are commonly used in literature in order to evaluate the different contributions of terms in the energy equation, as shown in a work by Doki-Thonon *et al.* [85] or in the study of turbulent phenomena in channel flows [95].

In order to evaluate the maximum impact of each term of the momentum equation 2.3b, one can compare the absolute value of the terms cell by cell after having solved the whole set of equations 2.3. The role of the single terms of the momentum equation is checked for a wide range of operational parameters in order to investigate a comprehensive range of Reynolds number. In the current analysis, the Reynolds number is computed through the definition 2.6 using the gap height H at the center of the pin surface as reference length. Different gap heights $1 < H < 20$ μm as well as different sliding velocities $0.01 < V < 1$ m/s were considered, leading to a range of the investigated Reynolds numbers which goes from $Re = 0.005$ to $Re = 10$. Density and viscosity were kept fixed to typical experimental values form [GMB$^+$15], namely $\rho = 936.3$ kg/m^3 and $\mu = 0.0187$ Pa·s.

Figure 2.11 shows the absolute value of the three terms of the momentum equation, namely the pressure gradient, the viscous term and the nonlinear term. The portrayed case represents one of the strictest condition, since its Reynolds number is $Re = 5$. In order to ease the visualization of the balance analysis, which is performed in the whole domain, the terms are visualized on the cutting plane passing through the half of the channel. As one can see, the pressure term and the viscous one have an almost coincident distribution and also very close maximal values, whilst the nonlinear term is in average three order of magnitude smaller. Therefore, one can conclude that the nonlinear terms are less relevant than the viscous and the pressure ones for what concerns the hydrodynamic flow in the macroscopic scales of the untextured pin.

2.3.3 Inertial effects in the texture scale

In the last decades, the Reynolds equation has been by far the most used mathematical framework in the numerical analysis of surface textures [16]. However, there exist some limitations in the applicability of such an equation due to the hypotheses considered in its derivation. These limitations are investigated in detail in the work of Dobrica and Fillon [46] concerning the hypotheses of the low Reynolds numbers and domain aspect ratio (length/height) of the considered geometry. In their work, these authors focused on a 2D rectangular

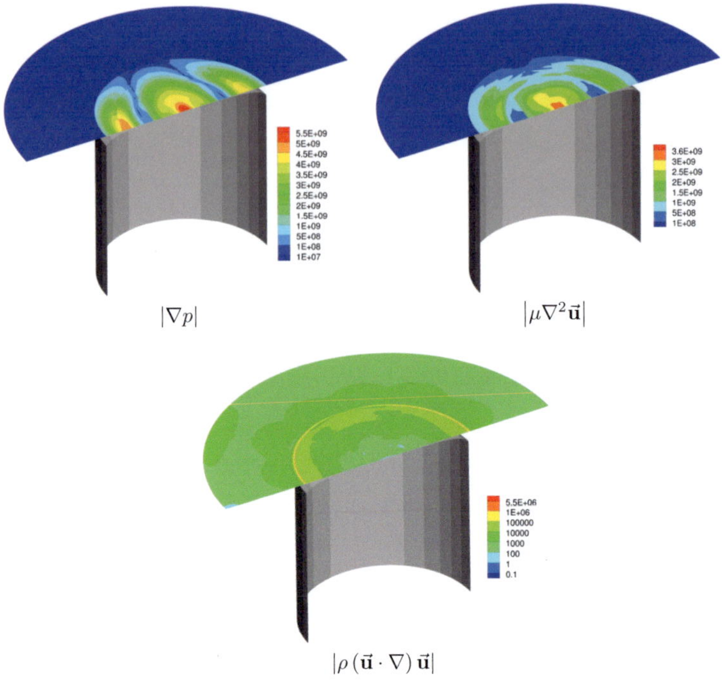

Figure 2.11: Contour of the magnitude of the pressure, viscous and nonlinear terms evaluated in a plane aligned with the centerline of the gap. The above disc (not shown) slides with a uniform linear velocity U from left to right. Due to the symmetry of the problem, only one half of the pin is considered. The y-axis is magnified 40 times. The prescribed flow parameters are: $U = 1$m/s, $H = 10$ μm, $\rho = 936.3$ kg/m^3 and $\mu = 0.0187$ Pa·s, giving a Reynolds number of $Re = 0.5$.

cavity (2D dimple) and compared the pressure distribution obtained with the Navier-Stokes equation to the one obtained with the Reynolds. The comparison was carried out for a wide range of Reynolds numbers and aspect ratios, the latter is defined as follows:

$$\lambda = \frac{l_D}{h_D} \tag{2.30}$$

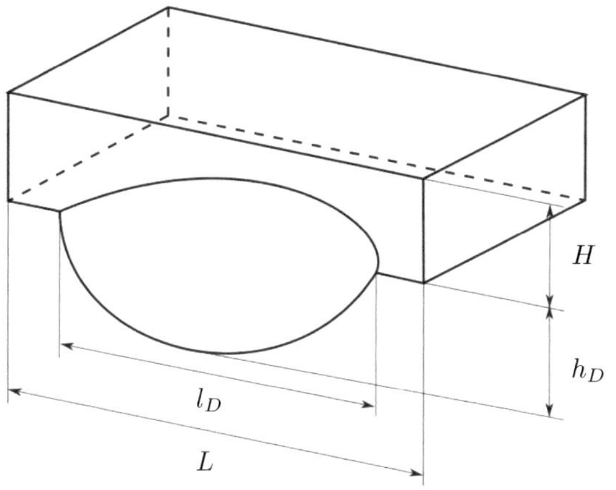

Figure 2.12: Sketch of the considered dimple geometry. In order to ease the comparison of the results, all the reference lengths are set consistently to those defined by Dobrica and Fillon in [46].

where l_D and h_D are the length and the depth of the dimple, respectively (as defined in figure 2.12).

The pressure difference between the Navier-Stokes and the Reynolds equation can be studied as a function of the Reynolds number and the aspect ratio λ by defining the pressure error Δ_1 as follows:

$$\Delta_1 = \int_A \left(\frac{|p_{Re}(x,z) - p_{NS}(x,z)|}{p^+_{averageNS}} \right) dxdz. \tag{2.31}$$

This formulation of pressure error is defined through the integration over the area A which corresponds, by convention, to the upper moving surface of the computational domain as shown in figure 2.12. Therefore, the definition 2.31 is the extension for 3D texture elements of the one proposed by Dobrica and Fillon in [46] which was originally presented only for 2D rectangular dimples.

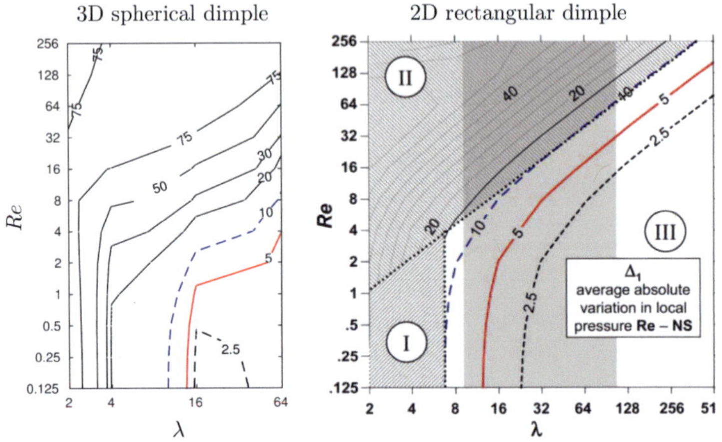

Figure 2.13: Comparison between the contour of the pressure error Δ_1 adapted from [Fur16] and the contour plot presented by Dobrica and Fillon in [46]. The considered aspect ratios on the left are $\lambda = 2, 4, 16, 64$.

In particular, this section shows the results concerning 3D spherical texture elements, since all the experimental campaigns mentioned in section 2.1 and in chapter 4 dealt with such a texture shape. Further analysis with other texture shapes can be found in the theses of Oscar Furst [Fur16] and Jingzhe Shu [Shu16].

The pressure field is computed with both the incompressible Navier-Stokes equations 2.4 (solved through the commercial solver FLUENT) and the incompressible version of the Reynolds solver presented in section 2.2.2. Both numerical methods employ periodic boundary conditions at the inlet and at the outlet where the reference ambient pressure is prescribed. The flow is driven by the movement of the upper wall with velocity V. Thanks to the symmetry of the geometry, only half of the domain is considered.

Figure 2.13 shows the contour of the pressure error Δ_1 obtained in collaboration with Furst [Fur16] and the contour distribution presented by Dobrica and Fillon in [46]. As one can see, the pressure error in case of a 3D spherical dimple presents a smaller applicability region with respect to the Reynolds number, meaning that the pressure distribution obtained with the Reynolds equation over a 3D spherical dimple is reliable for a smaller range of Reynolds numbers. A possible explanation could be the fact that the flow detachment at the leading edge of the 3D spherical dimple occurs for higher Reynolds numbers than in the case of 2D rectangular dimples. Even though one might deduce that the Δ_1 should be lower for a relatively smoother geometry such as the spherical dimple, streamline visualizations clearly pointed out that a sharp detachment can actually lead to a smaller difference between the Reynolds and the Navier-Stokes solutions. This may happen because the flow outside the dimple is less influenced by what happens inside when the detachment line is more prominent. Further, it has to be noted that the computation of Δ_1 may be more sensitive in the 3D case because the integration is extended to an additional dimension, hence making the computation of Δ_1 more sensitive to the pressure differences.

On the other side, the limit of the applicability region with respect to the aspect ratio λ remains with good approximation unvaried. As reported also by Dobrica and Fillon as practical guideline, the aspect ratio should not exceed the limiting value of $\lambda = 10$ in order to assure the applicability of the Reynolds equation. This limitation will be considered also in chapter 4 in which the extensive parametric study about surface textures is presented.

3 Relevance of viscous dissipation effects

This chapter focuses on the thermodynamic representation of lubricant flows with the aim to investigate the role of the viscous dissipation in thin film lubrication.

Even though the thermodynamic analysis of lubricant flows represents an important branch of the tribological research, the possibility to numerically predict the distribution of all thermodynamic variables, such as pressure, temperature and density in typical tribological applications still represents a challenge [96]. Since the first works on thermo-elastohydrodyamic lubrication (TEHL), it is well known that even small variations of one of these variables may drastically influence the other two leading to very different tribological performances of the analyzed device. In particular, the impact of viscosity and temperature on the pressure is referred to as viscosity wedge or thermal wedge in the works of Cameron [97] and Cope [98]. These effects may represent the only way to generate a non-zero pressure distribution if a simple lubricant shear flow between two parallel plates is considered. On the other side, they may lead to additional pressure losses when more complex systems are considered, especially when these effects are induced by viscous dissipation mechanisms [82, 83, 99]. More generally, the impact of such variations in the thermodynamic state of the system can imply both detrimental as well as positive consequences, as presented, for example, by Bruyere *et al.* in [67]. The underlying causes of pressure losses are, more commonly, a multidisciplinary matter of study which ranges from the thermo-elastohydrodynamic lubrication [100–102] to the investigation of micro channel applications in electronics and cooling devices [103, 104].

As underlined in the previous chapter, viscous dissipation is responsible for the strong coupling between the energy and the momentum equation (or the Reynolds equation, as shown in section 2.2). Through this coupling, the mechanical energy of the flow is typically transformed into internal energy, leading to a temperature increase. More generally, viscous dissipation induces a heat generation which, in turn, increases the temperature of the fluid. This has important consequences on the behavior of the system, since this mechanism is further enhanced by the the high sensitivity of the viscosity to temperature variations. Many studies showed that ignoring the heat generated trough viscous dissipation can result in a wrong estimation of important tribological quantities, such as the film thickness and the friction coefficient [105–107].

The implication of viscous dissipation in TEHL analyses were studied, among others, by Habchi *et al.* in [84], pointing out that compressibility and thermal effects are particularly enhanced by the severe operating conditions and the small high aspect ratio L/h of the geometry close to the contact region. The energy transfer mechanisms can also be studied by means of a simple energy balance of heat fluxes, as done by Doki-Thonon *et al.* in [85], showing that viscous dissipation represents the largest contribution among all thermal effects.

In the last decades, TEHL models have been constantly improved in order to describe the contact performance in complex systems and under critical conditions. Nonetheless, only qualitative relationships have been proposed between certain operating parameters and the corresponding changes in the thermodynamic state of the system [66, 103, 108]. For this reason, this chapter focuses on quantitative scaling laws for viscous dissipation with respect to the flow parameters and to the lubricant properties.

Both the compressible Navier-Stokes equations and the thermal-compressible Reynolds equation presented in section 2.2 are employed in this chapter with the aim to quantitatively evaluate the dependence of the flow parameters on the pressure losses and the thermal increase . Furthermore, the relevance of each term of the energy equation is investigated by means of an energy balance analysis, providing a corresponding scaling law with respect to the flow conditions. Part of the scaling laws presented in this chapter are also presented in one of the author's publications [CSMF17].

3.1 Problem definition

The numerical set-up considered in this chapter describes the steady flow of a compressible lubricant fluid sheared in a 2D channel whose height h is constant. Such a numerical set-up is representative of typical tribological applications and, at the same time, simple enough in order to isolate the effects caused by the viscous dissipation from those induced by the complexity of the geometry [48, 109].

Figure 3.1 shows the plane channel geometry which is considered in this chapter. The boundary conditions are assigned in a consistent way for both the chosen numerical methods, namely the compressible Navier-stokes equation and the thermal-compressible Reynolds equation. The flow is driven by the upper surface which moves at the velocity U (where $U = \vec{V}_{up} \cdot \hat{x}$) while the lower one is fixed; both walls are adiabatic. At the inlet, the static temperature T_{in} and zero pressure gradient $\left.\frac{dp}{dx}\right|_{x=0} = 0$ are prescribed together with a linear velocity profile. At the outlet, static pressure p_{out} is assigned while all other variables have a zero second derivative. These conditions allow the temperature distribution and the velocity field to freely adapt at the outlet, hence having less invasive boundary conditions. The same is valid at the inlet, where the pressure value can adapt according to the pressure gradient required to balance the loss. In this way, the pressure loss can be measured as $\Delta p = p_{in} - p_{out}$. A schematic representation of the pressure loss and the increase of temperature is also proposed in figure 3.1.

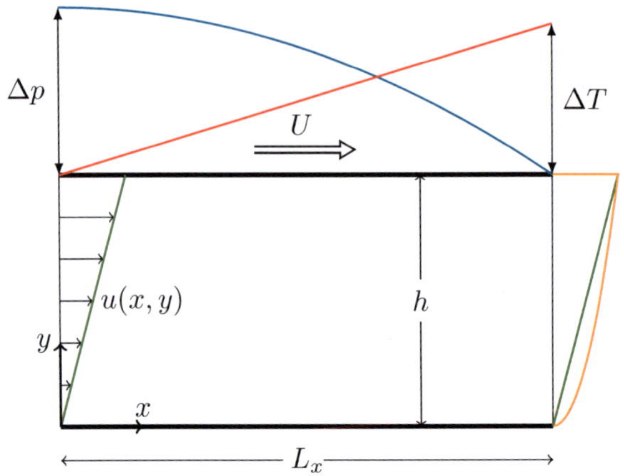

Figure 3.1: Schematic representation of the flow between the two parallel walls. The channel has height h and length L_x. The flow is driven by the movement of the upper wall whose velocity is U. The pressure loss is measured as the difference between the inlet and the outlet pressure $\Delta p = p_{in} - p_{out}$. In the same way, the temperature increase is defined as $\Delta T = T_{out} - T_{in}$. A Couette velocity profile is prescribed at the inlet. This sketch also portrays the two different kind of velocity profile which can occur at the outlet, namely a linear one (Coutte) or a parabolic one (Poiseuille).

A mineral base oil is chosen as lubricant for this analysis. The oil properties are taken from the experimental characterization carried out by Regueira *et al.* in [110]. As shown in section 2.2, two equations of state (EOS) are considered in this study, namely the stiffened gas EOS [111] and the Dowson-Higginson EOS in the form presented in [110]. The stiffened gas EOS is used in the first part of this analysis when the validity of the thermal-compressible Reynolds equation is discussed by comparing it to the compressible Navier-stokes equation. This is due to the fact that the Navier-Stokes solver includes only the stiffened gas as possible equation of state for compressible liquids. After this validation step, the thermal Reynolds solver is run only with the Dowson-Higginson equation of state. Further details about the implementation of the equation of state are shown in appendix A. It is important to notice that, although the Dowson-Higginson EOS fits better the experimental data

[110], the results presented in this chapter show a very weak dependence on the kind of EOS which is chosen [CSMF17].

The computational grid for the simulations with the compressible Navier-Stokes equation consists of a 2D mesh with 256 points in the streamwise direction x and 64 points in the wall normal direction y. The results are proved to be mesh-independent through the application of 5 multi grid levels, so that there is a negligible difference in the solution between the last two multi grid levels. The steady solution of every multi grid level is computed in the pseudo time until the relative residuum drops below a certain tolerance, namely 10^{-5}. The computational grid for the thermal-compressible Reynolds approach consists of a 1D mesh with 256 points.

3.2 Validity of the thermal-compressible Reynolds approach

In this section the comparison between the compressible Navier-Stokes equations and the thermal-compressible Reynolds equation is presented. This analysis is needed in order to validate the approach with the Reynolds equation and to quantitatively identify the parameter range in which this approach can be safely applied. Once assessed the validity region, the computational advantages of the thermal-compressible Reynolds equation can be fully exploited to carry out an extensive parametric study with the aim of investigating the effects of viscous dissipation.

The main differences between the two approaches rely on the two main hypotheses which are made in the derivation of the Reynolds model. The first one is that the inertial terms are considered negligible in the Reynolds model, while the second one concerns the assumption of constant flow properties in the wall normal direction y. Three non-dimensional numbers can be used to describe the applicability of these two hypotheses. The first one is the Reynolds number defined using the channel height as presented in section 2.2.2:

$$Re = \frac{\rho U h}{\mu}. \tag{3.1}$$

The second non-dimensional number is the Péclet number (recalled from equation 2.20):

$$Pe = \frac{\rho C_p U h}{k_T} \tag{3.2}$$

which expresses the ratio between the heat convected by the fluid motion and the heat diffused through the fluid itself. The Péclet number is also directly related to the temperature distribution in thin-film lubrication, in particular, at small Péclet values the temperature results constant in the wall normal direction [66]. The third non-dimensional number is the Sommerfeld number (recalled from equation 2.20):

$$\mathcal{S} = \frac{\mu U L_x}{h^2 p_r} \tag{3.3}$$

where the reference pressure p_r is set equal to the outlet pressure $p_r = p_{out}$ for the whole analysis in this chapter. As shown further in section 3.4, the Sommerfeld number is also the scaling factor which multiplies the viscous dissipation term in the energy equation 2.16 when the latter is brought into the dimensionless formulation.

To investigate the validity of the approach with the Reynolds model, the impact of the two above-mentioned hypotheses has to be highlighted. The requirement of low Reynolds number is typically easily fulfilled, due to the channel geometry and the laminar nature of the flow in typically tribological applications [78]. On the other hand, the hypothesis of constant flow properties across the channel is more strict and requires a more careful treatment. For small Sommerfeld numbers, this approximation holds rather well, since the viscous dissipation term in the energy equation remains almost negligible and no other possible heating-up mechanism can change the thermodynamic status of the flow. Therefore, the channel temperature remains constant as well as the density and the pressure distributions. In this case, the channel flow fulfills the so-called Couette solution and maintains its linear velocity profile unchanged from the inlet to the outlet.

Conversely, the behavior of the flow changes drastically when the Sommerfeld number is high enough to trigger the heating of the fluid through viscous dissipation. Thanks to this generated heat, the temperature increases along the channel implying that the density and the viscosity have to decrease according to the equation of state and the viscous laws. At this point, the mass conservation implies that the velocity profile has to change in order to contrast the decrease of the density. The velocity profile evolves, hence, from the linear one prescribed at the inlet to a parabolic one at the outlet. This corresponds to the generation of a streamwise pressure gradient. Therefore, the flow becomes shear- and pressure-driven, assuming the typical Couette-Poiseuille configuration.

The formation of a pressure gradient is related to a dissipation phenomenon and has to be considered as an additional source of mechanical energy from the outside (*e.g.* pumping power) in order to cope with the pressure loss due to the internal dissipation of the system. Such an occurrence of pressure loss is then associated to the so-called "thermal wedge", since the variations in the pressure distribution are driven only by changes in the thermodynamic state and not by changes of the geometry ("geometry wedge") [97, 98].

The presence of a pressure gradient also affects the shear stress distribution in the channel, which is defined as follows:

$$\tau_{yx} = \mu \frac{du}{dy} = \frac{1}{2} \frac{dp}{dx} (2y - h) + \mu \frac{U}{h}. \tag{3.4}$$

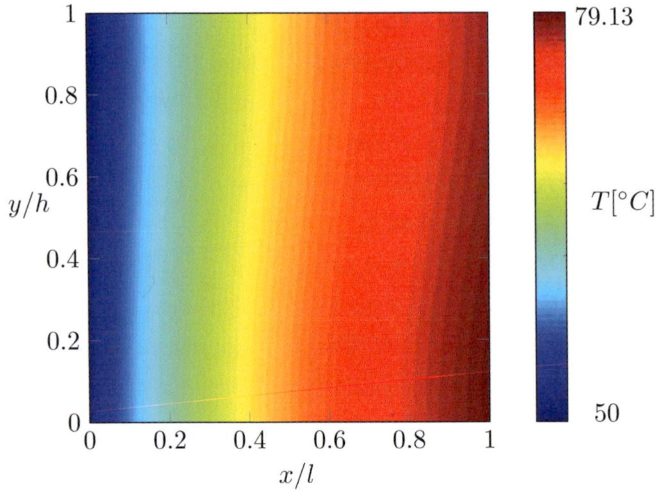

Figure 3.2: Temperature distribution in the channel computed with the compressible Navier-Stokes equation. The simulation is carried out with the following non-dimensional numbers: $\mathcal{S} = 590.2$, $Re = 1.1442$, $Pe = 4636.08$. Adapted from [CSMF17].

Due to $\frac{dp}{dx} < 0$, the pressure term of the shear stress results to be positive on the lower wall and negative on the upper wall. Since the second term $\mu \frac{U}{h}$ is always positive, the shear stress is higher at the lower wall than at the upper one. This leads to higher viscous dissipation in the lower part of the channel and, consequently, a temperature gradient in the wall normal direction can appear.

Simulation results at high Sommerfeld number reveal, indeed, a higher temperature in the lower part o the channel. Figure 3.2 shows the temperature distribution as computed with the compressible Navier-Stokes equations for a representative case with $\mathcal{S} = 590.2$, $Re = 1.1442$ and $Pe = 4636.08$. In this simulation, the maximum temperature difference between the lower and the upper wall is $\Delta T_y = 1.38$ [°C].

The wall normal temperature gradient cannot be caught by the numerical approach based on the thermal-compressible Reynolds equation. For this reason, further simulations were carried out with different Reynolds and Péclet

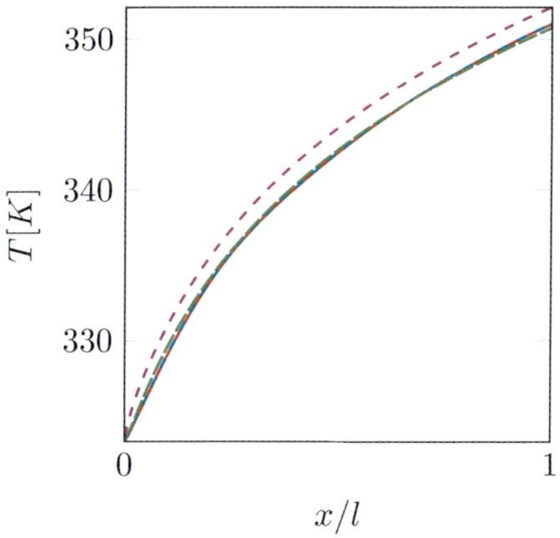

Figure 3.3: Temperature distribution along the x-axis obtained with the compressible Navier-stokes equation. Two simulations with the same Sommerfeld number $\mathcal{S} = 590.2$ but different Péclet and Reynolds numbers are shown. Upper (——) and lower (- - -) wall temperature for the case with $Re = 0.035$, $Pe = 146.84$. Upper (—–-) and lower (- - -) wall temperature for the case with $Re = 1.1442$, $Pe = 4636.08$. Adapted from [CSMF17].

numbers. Figure 3.3 shows the temperature distribution at both walls for two different cases with the same Sommerfeld number but with different Reynolds and Péclet numbers. As one can see, for the same value of Sommerfeld number, the wall normal temperature gradient occurs only when a certain threshold in the Reynolds and Péclet numbers is exceeded. More generally, it was found that the hypothesis of constant flow properties across the channel is valid for values of $Pe < 10^3$ and $Re < 0.1$. Therefore, the application of the thermal-compressible Reynolds model as formulated in equations 2.10 and 2.16 will be restrained to this range. An extension of this limitation could be obtained by considering the generalized Reynolds equation, as presented by Dowson in [78], since this model can take into account also the variation of the flow properties across the channel.

As further validation of the thermal compressible Reynolds model, figures 3.4 and 3.5 compare the pressure, density and temperature distributions computed

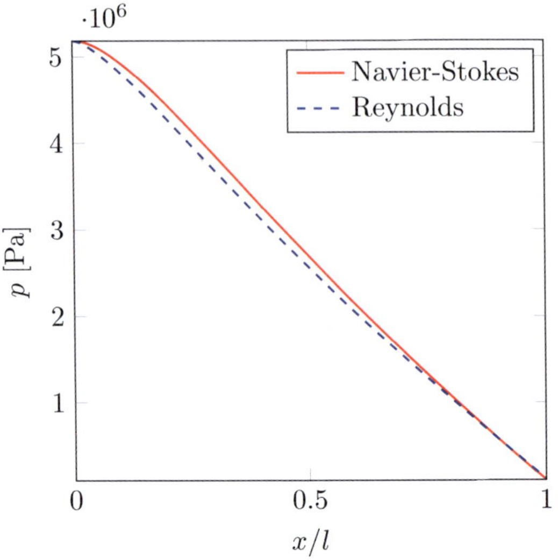

Figure 3.4: Comparison of the pressure distribution along the non-dimensional streamwise direction x/L_x obtained with the Navier-Stokes equations and the thermal-compressible Reynolds equation. The simulation is carried out with $\mathcal{S} = 590.2$, $Re = 0.035$, $Pe = 146.84$. Adapted from [CSMF17].

with both models. The figures show the results obtained with a Sommerfeld number of $\mathcal{S} = 590.2$ and two Péclet and Reynolds numbers which satisfy the above-mentioned applicability condition, namely $Re = 0.035$ and $Pe = 146.84$. As one can see, the results show a good agreement between both numerical models. This allows to exploit the higher computational efficiency of the Reynolds model in order to further investigate the impact of viscous dissipation throughout this chapter.

It has to be noticed that the results in this section were obtained by coupling both models with the stiffened gas EOS, while in the following sections the Dowson-Higginson EOS will be used.

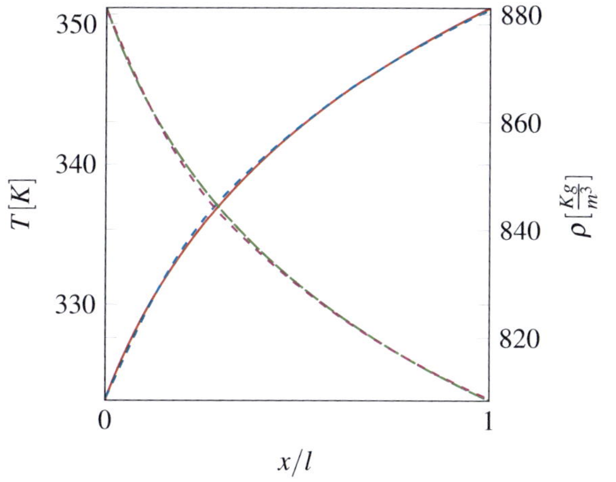

Figure 3.5: Comparison of the temperature and density distributions along the non-dimensional streamwise direction x/L obtained with the Navier-Stokes equations and the thermal-compressible Reynolds equation. Temperature (——) Navier-Stokes, temperature (- - -) Reynolds. Density (—–-) Navier-Stokes, density (- - -) Reynolds. These results are obtained at $\mathcal{S} = 590.2$, $Re = 0.035$, $Pe = 146.84$. Adapted from [CSMF17].

3.3 Scaling of pressure losses and temperature increase

As shown in the previous section, as the Sommerfeld number increases, the effects related to viscous dissipation become more evident and induce important variations in the distribution of the thermodynamic variables, such as pressure and temperature. The impact on pressure and temperature can be quantified by considering the variation between the inlet and the outlet of the channel. In particular, since the pressure variations occur due to a dissipation mechanism, we can define the pressure loss as $\Delta p = p_{in} - p_{out}$ as shown in figure 3.1. Similarly, the temperature increase is defined as $\Delta T = T_{out} - T_{in}$.

In order to understand the behavior of the pressure loss Δp and the temperature increase ΔT under different operating conditions, an extensive parametric study is carried out based on the four flow parameters which characterize

$1 < h < 100$	$[\mu m]$
$10^{-6} < l < 10^{-3}$	$[m]$
$7.9 \cdot 10^{-2} < \mu_0 < 7.9 \cdot 10^{-5}$	$[Kg/(ms)]$
$10^{-4} < U < 1$	$[m/s]$

Table 3.1: Range of the four flow parameters used to compute the Sommerfeld number $\mathcal{S} = \frac{\mu_0 l U}{h^2 p_{ref}}$. Adapted from [CSMF17].

the channel flow, namely the gap height h, the channel length L_x, the upper wall velocity U and the viscosity μ. These four parameters constitute the Sommerfeld number. Therefore, the dependence of Δp and ΔT can be studied with respect to the Sommerfeld number, whereas the reference pressure p_r used in the definition 3.3 can be set to the outlet pressure, since it represents a constant for all the simulations.

Figures 3.6 and 3.7 portray the behavior of the normalized pressure loss and temperature increase as function of the Sommerfeld number. Each of the points represented in the figures corresponds to a single simulation at a specific value $\mathcal{S}$ which is obtained through different combinations of the four parameters h, L_x, U and μ. The investigated parameter range is presented in table 3.1. Particular attention was paid to the setting of the parameter range in order to fulfill the applicability condition for the thermal-compressible model as expressed in the previous section ($Pe < 10^3$ and $Re < 0.1$).

Figure 3.6 shows that a correlation exists between the normalized pressure loss $\frac{\Delta p}{p_r}$ and the Sommerfeld number $\mathcal{S}$. Furthermore, for Sommerfeld numbers below $\mathcal{S} < 10$ the normalized pressure loss scales with the square of the Sommerfeld number $\frac{\Delta p}{p_r} \propto \mathcal{S}^2$. Also the temperature increase is found to scale with the Sommerfeld number, but in a linear way: $\frac{\Delta T}{T_r} \propto \mathcal{S}$. This denotes the role of the Sommerfeld number as a key parameter in lubrication theory also for what concerns the effects of viscous dissipation and their implication on the pressure and temperature distributions. This scaling can be successfully

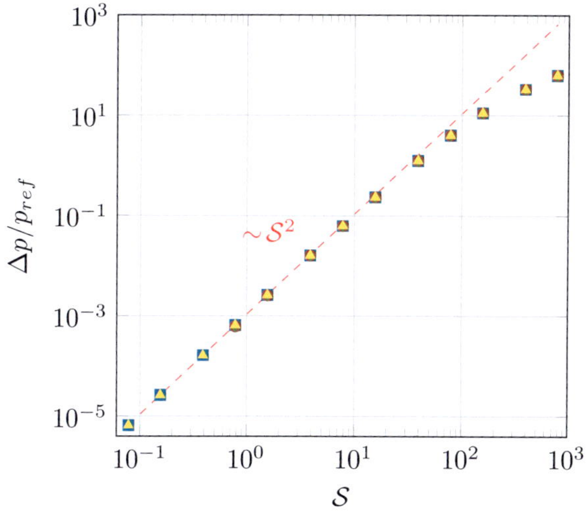

Figure 3.6: Normalized pressure losses $\Delta p/p_{out}$ as a function of the Sommerfeld number $\mathcal{S}$ which is obtained through different combinations of the four flow parameters (■) h, (●) L_x, (♦) U, (▲) μ. The fitted dashed line marks the linear proportionality of the temperature increase with the Sommerfeld number. Adapted from [CSMF17].

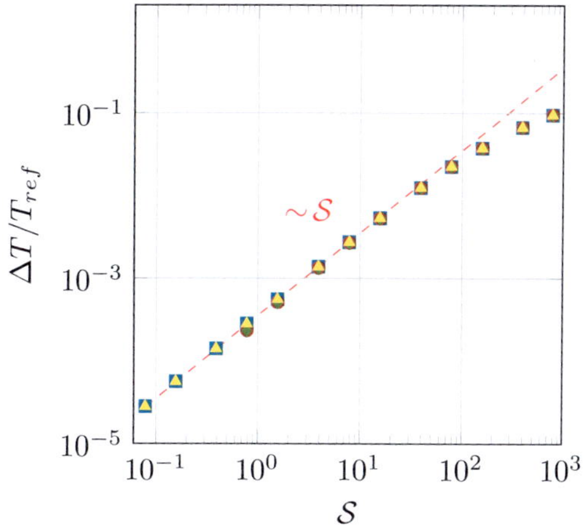

Figure 3.7: Normalized temperature increase $\Delta T/\Delta T_{out}$ as a function of the Sommerfeld number $\mathcal{S}$ which is obtained through different combinations of the four flow parameters (■) h, (●) L_x, (♦) U, (▲) μ. The fitted dashed line marks the linear proportionality of the temperature increase with the Sommerfeld number.

used for the quantitative estimation of the pressure loss and temperature increase. For example, by doubling the viscosity value, one can know *a priori* that the pressure loss will be four times higher, while the temperature at the outlet will be twice as much as at the inlet.

The behavior of Δp and ΔT presented here are obtained in the reference scenario of a plane channel flow. Therefore, a quantitative extension in a more complex geometry cannot directly be done because of the possible presence of "geometry wedge" effects [94].

3.4 Energy balance analysis

The heat transfer mechanisms in the flow can be analyzed by means of a decomposition of the energy equation, in order to understand the role of each term in the energy balance. Similar approaches are common in literature and have been used, for example, by Doki-Thonon *et al.* in [85] for the analysis of the heat fluxes in spinning contacts or by Huang *et al.* for the study of compressible turbulent flow [112]. The aim of this analysis is to evince the underlying scaling between the the single term of the energy equation and the operating parameters. The dependence on the latter can be subsequently generalized with the aid of the Sommerfeld number.

The energy balance analysis can be performed by first integrating the energy equation 2.3c over the whole domain $\mathcal{V}$ of the channel

$$\int_{\mathcal{V}} \nabla \cdot \left((\rho e^t + p) \cdot \mathbf{u} \right) d\mathcal{V} = \int_{V} \nabla \cdot (k \nabla T + \mathbf{u} \cdot \sigma) \, d\mathcal{V}. \tag{3.5}$$

By applying the divergence theorem and by assuming a unitary width of the channel, the above equation can be rewritten as follows:

$$\begin{gathered} \int_0^{h_{out}} (\rho e^t + p) \mathbf{u} dy - \int_0^{h_{in}} (\rho e^t + p) \mathbf{u} dy = \\ \int_0^{h_{out}} k \nabla T \hat{n} dy - \int_0^{h_{in}} k \nabla T \hat{n} dy + \int_0^{h_{out}} \mathbf{u} \cdot \sigma \hat{n} dy - \int_0^{h_{in}} \mathbf{u} \cdot \sigma \hat{n} dy. \end{gathered} \tag{3.6}$$

In this way, the integral formulation of the energy equation expresses the conservation of the total enthalpy over the whole channel domain per unit of width [W/m]. The normal versor $\hat{n}$ is assumed to be positive in the outward direction. The integrals are evaluated at the channel inlet and outlet.

The terms of equation 3.6 can be regrouped and renamed in order to ease the discussion of the balance analysis. The terms on the left hand side represent the net enthalpy flux along the channel:

$$\Delta W_{H^t} = \int_0^{h_{out}} (\rho e^t + p)\mathbf{u}dy - \int_0^{h_{in}} (\rho e^t + p)\mathbf{u}dy = W_{H^t_{out}} - W_{H^t_{in}}. \tag{3.7}$$

The first term on the right hand side in equation 3.6 corresponds to the net power diffused by the thermal conduction along the channel and can be summarized as follows:

$$\Delta W_{T_{diff}} = \int_0^{h_{out}} k\nabla T\hat{n}dy - \int_0^{h_{in}} k\nabla T\hat{n}dy. \tag{3.8}$$

Finally, the power dissipated by the viscous dissipation is renamed as

$$\Delta W_{diss} = \int_0^{h_{out}} \mathbf{u}\cdot\sigma\hat{n} - \int_0^{h_{in}} \mathbf{u}\cdot\sigma\hat{n}. \tag{3.9}$$

Therefore, equation 3.5 can be expressed in the following short form:

$$\Delta W_{H^t} = \Delta W_{T_{diff}} + \Delta W_{diss}. \tag{3.10}$$

Due to the hypothesis of adiabatic walls, no heat exchange terms are considered in equation 3.10. Equation 3.10 can be physically interpreted as the balance between the net enthalpy flux ΔW_{H^t} and two contributions consisting of the power diffused by thermal conductivity of the fluid $\Delta W_{T_{diff}}$ and the power dissipated by viscosity ΔW_{diss}.

The enthalpy flux ΔW_{H^t} can be further decomposed, in order to distinguish between the contributions of the mechanical energy and the internal energy (*i.e.* thermal energy). In this way, the net flux of the mechanical energy becomes

$$\Delta W_{E_i} = \int_0^{h_{out}} (p)\mathbf{u}dy - \int_0^{h_{in}} (p)\mathbf{u}dy \tag{3.11}$$

while the net flux of the internal energy reads

$$\Delta W_{E_i} = \int_0^{h_{out}} (\rho e^t)\mathbf{u}dy - \int_0^{h_{in}} (\rho e^t)\mathbf{u}dy. \tag{3.12}$$

Therefore, the net enthalpy flux is decomposed as:

$$\Delta W_{H^t} = \Delta W_{E_i} + \Delta W_{E_m}. \tag{3.13}$$

This decomposition will be extensively used in the following discussion in order to highlight the mechanism related to the pressure loss and the temperature increase.

The flow in the channel is driven by the movement of the upper wall, which consequently represents the only source of energy in the channel. The power introduced by the upper wall can be expressed by integrating the wall shear stress times the velocity of the upper wall U.

$$W_{wall} = \int_{S_{wall}} \tau_{yx} U dxdz. \tag{3.14}$$

By considering a global balance of channel, as schematically portrayed in figure 3.8, one can see that the power W_{wall} introduced by the movement of the wall has to be balanced by the net enthalpy flux ΔW_{H^t} of the entire channel:

$$\Delta W_{H^t} = W_{wall}. \tag{3.15}$$

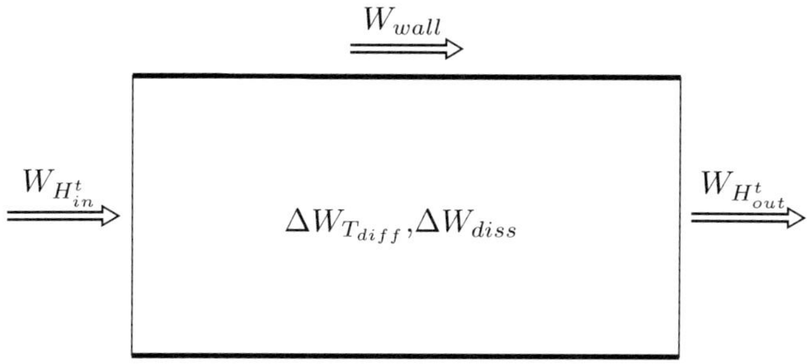

Figure 3.8: Visual representation of the energy balance expressed in equation 3.16. The net flux of power ΔW_{H^t} is computed as the outlet-inlet difference $\Delta W_{H^t} = W_{H^t_{out}} - W_{H^t_{in}}$. The moving wall introduces the power W_{wall} in the system, while the flow in the channel diffuses the power $\Delta W_{T_{diff}}$ and dissipates the power ΔW_{diss}. Adapted from [CSMF17]

The equivalence expressed in equations 3.10 and 3.15 can be, hence, regrouped together in order to highlight the relationships between all the energy balance terms analyzed so far:

$$W_{wall} = \Delta W_{H^t} = \Delta W_{T_{diff}} + \Delta W_{diss}. \tag{3.16}$$

The behavior of the energy balance terms outlined in equation 3.16 can be studied as function of the operating conditions by means of a parametric study as done in the previous section for the pressure loss and the temperature increase. The considered parameter space concerning L_x, h, U and μ is shown in table 3.1. The simulations are carried out with the thermal-compressible Reynolds approach and particular attention is posed in the fulfillment of the applicability region in terms of Reynolds and Péclet numbers as shown in section 3.2. The terms of the energy balance are computed in the post-processing using the same resolution of the simulations and considering 100 points in the wall normal direction. In this way, the fulfillment of equation 3.16 was always verified with a maximal error margin of 0.3%.

Similar to what was done in section 3.3 for the pressure loss, the terms of equation 3.16 and 3.13 are analyzed as a function of the Sommerfeld number

$\mathcal{S}$. Figure 3.9 shows the six contributions that we can consider out of equation 3.10 and 3.15, namely the net flux of enthalpy ΔW_{H^t}, the net flux of internal energy ΔW_{E_i}, the power diffused by heat conduction $\Delta W_{T_{diff}}$, the power dissipated by viscosity ΔW_{diss}, the power introduced by the moving wall ΔW_{wall} and the loss of mechanical power $|\Delta W_{E_m}|$. As one can see, the terms corresponding to ΔW_{H^t}, ΔW_{wall} and ΔW_{diss} overlap perfectly, because the contribution of the temperature diffusion $\Delta W_{T_{diff}}$ is almost negligible at low Péclet and Reynolds numbers. Therefore, equation 3.16 can be simplified to:

$$W_{wall} = \Delta W_{H^t} = \cancelto{0}{\Delta W_{T_{diff}}} + \Delta W_{diss}. \tag{3.17}$$

This means that the power introduced by the wall is completely dissipated by the viscous dissipation and corresponds to the net increase of enthalpy along the channel. This finding is in agreement with the work of Doki-Thonon [85], in which the viscous dissipation is found to be the dominant physical phenomenon responsible of the temperature increase in the lubricated contact.

Thanks to the decomposition of the enthalpy in equation 3.13, one can also study the dependence of the internal and mechanical energy on the Sommerfeld number. As a result, the net flux of mechanical energy is always negative, confirming that the mechanical energy can only diminish in the system. For this reason the contribution of the mechanical energy is portrayed in figure 3.9 through its absolute value $|\Delta W_{E_m}|$. The loss of mechanical energy corresponds, on the other side, to an increase in internal energy ΔW_{E_i}. Moreover, the mechanical loss $|\Delta W_{E_m}|$ is almost negligible for small values of the Sommerfeld number $\mathcal{S}$, but they become more important as soon as $\mathcal{S} > 10$.

The behavior of the energy equation terms shown in figure 3.9 is obtained by varying only the channel height h inside the Sommerfeld number $\mathcal{S}$. The scaling relationship can be further generalized thanks to a non-dimensionalization of the energy equation 3.5. After carrying out a dimensional analysis, this

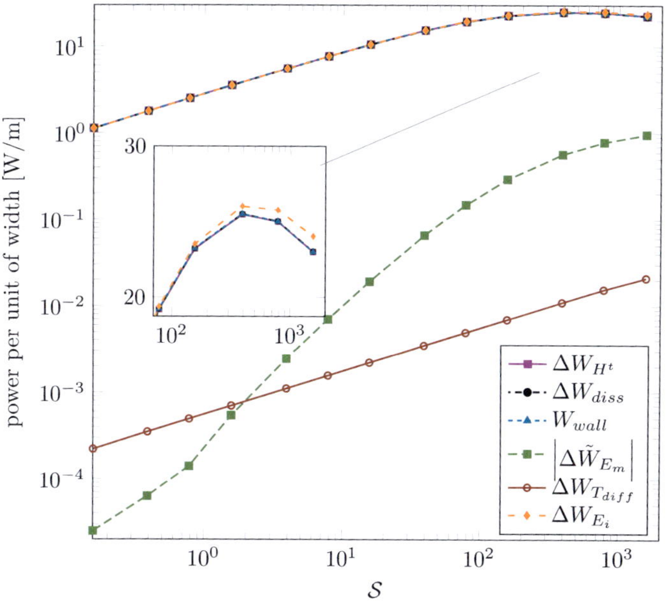

Figure 3.9: Contributions of the energy equation as function of the Sommerfeld number $\mathcal{S}$ (according to the decomposition done in the equations 3.7,3.8,3.9,3.10 and 3.14). Lines are introduced as guideline to the eye. Adapted from [CSMF17]

non-dimensionalization can be achieved by dividing equation 3.5 through the following dimensional group:

$$\beta = \frac{\mu U^2 L_x}{h}. \tag{3.18}$$

The behaviors of the so-obtained dimensionless terms of the energy equation are shown in figure 3.10 as a function of the Sommerfeld number. In these simulations, the Sommerfeld number is computed by varying all four operating parameters, namely L_x, h, U and μ. Also in this case the three terms W_{wall}, ΔW_{H^t} and ΔW_{diss} from equation 3.17 collide on the same curve. This is also a sign that the Sommerfeld number represents a good scaling parameter for the representation of the energy term, since different combinations of the four

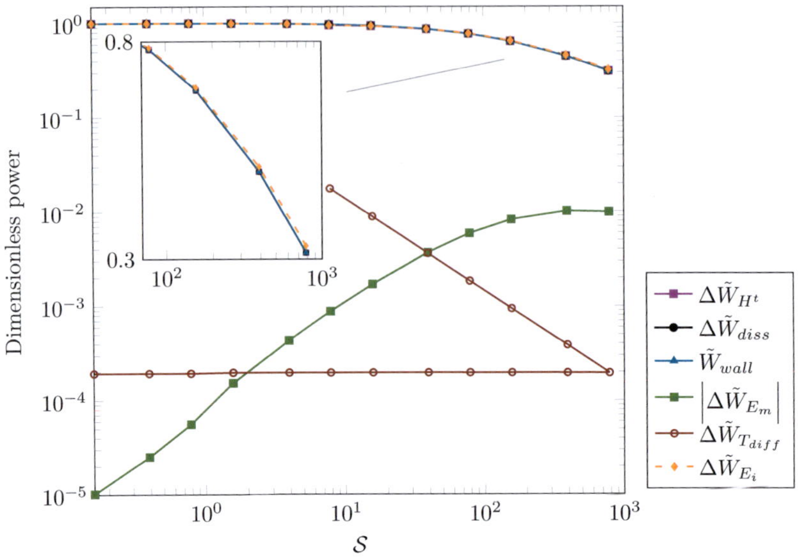

Figure 3.10: Dimensionless contributions of the energy equation as function of the Sommerfeld number $\mathcal{S}$ (according to the decomposition done in the equations 3.7,3.8,3.9,3.10 and 3.14). The non-dimensionalization is computed through division of the terms in figure 3.9 by the factor $\frac{\mu U^2 L_x}{h}$. Lines are introduced as guideline to the eye. Adapted from [CSMF17]

operating parameters lead to the same simulation results if the Sommerfeld number remains unvaried. Only the term corresponding to the temperature diffusion does not scale with the Sommerfeld number. This is due the fact that the temperature diffusion scales by definition with the Péclet number, which is why one sees two different lines in figure 3.10 for the term $\Delta W_{T_{diff}}$.

It is clear from figures 3.9 and 3.10 that an energy transfer mechanism occurs at high Sommerfeld numbers, which leads to a degradation of the mechanical energy into internal energy. This energy degradation mechanism consequently leads to a pressure loss and to an increase of temperature in the channel.

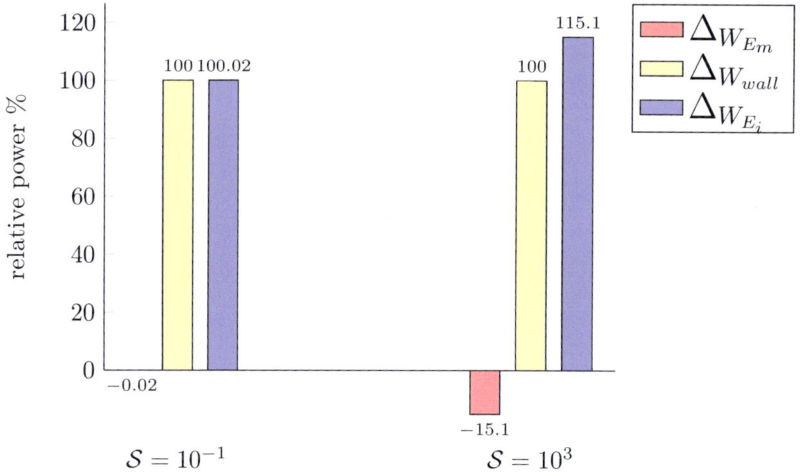

Figure 3.11: Histogram representation of the enthalpy decomposition as presented in equation 3.13. Two groups of histograms are shown at two different values of the Sommerfeld number. The height of the columns represents the relative, namely the respective value of W_{wall}, ΔW_{E_i} and ΔW_{E_m} normalized by net flux of enthalpy $\Delta W_{H'}$ (which is equal to the power introduced by the wall according to equation 3.17).

In order to portray more directly the impact of high Sommerfeld numbers on the thermodynamics of the flow, figure 3.11 shows a histogram representation of the enthalpy decomposition (equation 3.13). As one can see, for a small value of the Sommerfeld number, namely $\mathcal{S} = 0.1$, the mechanical loss is almost negligible and all the power introduced by the wall goes into the increase of the net flux of internal energy ΔW_{E_i}. Conversely, for a relatively high Sommerfeld number $\mathcal{S} = 10^3$, about 15% of the power introduced by the wall is subtracted to the mechanical energy and used to increase the internal energy of the system.

Role of the Sommerfeld number in the energy equation

The scaling laws of the effects related to viscous dissipation can be obtained also through a dimensional analysis of the energy equation itself. Indeed, since we know that the temperature diffusion is negligible at low Péclet numbers, one can recall the energy equation 2.3c in the following simplified form:

$$\nabla \cdot \left(H^t \cdot \mathbf{u}\right) = \nabla \cdot (\mathbf{u} \cdot \sigma). \tag{3.19}$$

This equation can be further simplified for the considered case of a 2D channel flow:

$$\frac{\partial}{\partial x}\left(H^t u\right) = \mu \left(\frac{\partial u}{\partial y}\right)^2. \tag{3.20}$$

In order to carry out a dimensional analysis, the following quantities have to be redefined in terms of non-dimensional and reference variables: $\frac{\partial}{\partial x} = \left(\frac{1}{l}\right)\frac{\partial}{\partial \tilde{x}}$, $\frac{\partial}{\partial y} = \left(\frac{1}{h}\right)\frac{\partial}{\partial \tilde{y}}$, $u = U\tilde{u}$, $\mu = \mu_0\tilde{\mu}$, $H^t = H_0^t\tilde{H}^t$, where $(\tilde{\cdot})$ refers to a dimensionless variable. These non-dimensional relationships can be substituted into equation 3.20:

$$\frac{1}{l}\frac{\partial}{\partial \tilde{x}}\left(H_0^t\tilde{H}^t U\tilde{u}\right) = \mu_0\tilde{\mu}\left(\frac{U}{h}\frac{\partial \tilde{u}}{\partial \tilde{y}}\right)^2. \tag{3.21}$$

By rearranging the terms and making the non-dimensional total enthalpy gradient explicit it follows that:

$$\frac{\partial}{\partial \tilde{x}}\tilde{H}^t = \frac{\mu_0 U l}{h^2 H_0^t}\tilde{\mu}\left(\frac{\partial \tilde{u}}{\partial \tilde{y}}\right)^2 \tag{3.22}$$

where the factor on the left hand side can now be re-elaborated as a function of the Sommerfeld number in the following way:

$$\frac{\partial}{\partial \tilde{x}}\tilde{H}^t = \frac{S p_r}{H_0^t}\tilde{\mu}\left(\frac{\partial \tilde{u}}{\partial \tilde{y}}\right)^2 \tag{3.23}$$

where both p_r and H_0^t can be arbitrarily chosen as constant.

Equation 3.23 shows how the Sommerfeld number corresponds to the scaling factor between the viscous dissipation and the enthalpy gradient in the streamwise direction. This is in agreement with the numerical results shown in the previous figures, where the viscous dissipation effects get higher at high values of the Sommerfeld number. As a more general consideration, at high values of Sommerfeld number the viscous dissipation mechanisms increase the amount of power which is degraded from mechanical to internal power, hence lowering the efficiency of the system. This could be the case for high performance applications, where the small size and the high load typically lead to high Sommerfeld numbers.

4 Optimal texture shape and position

In this chapter we want to apply the numerical models, which were developed and tested in the previous part of this work, with the aim to study the effect of surface textures on the tribological performance of a pin-on-disc tribometer.

The relevance of surface texturing and the interest which it aroused in the scientific community were already highlighted in the introduction of this work (see section 1.1). In the following, the focus of the literature survey is oriented to the works which show closer analogies to the investigation presented in this chapter. In particular, the numerical analysis presented in this chapter is closely related to three experimental campaigns which were carried out at the institute of applied material IAM-CMS at the Karlsruhe Institute of Technology KIT.

Among them, the most relevant experimental campaign concerns the role of viscosity in the determination of the texture design parameters which lead to the minimum friction coefficient. Many studies in the past focused on the search and determination of the optimal texture parameters, nonetheless consensus has not been achieved universally yet in the research community. This is mostly due to the big variety of investigation methods which were employed and the large arbitrariness in which texture design parameters can be chosen. For example, in case of textures made of spherical dimples, many experimental studies based on pin-on-disc tribometers found contradicting optimal aspect ratios of dimples (*i.e.* the ratio between the length and depth of a dimple $\lambda = \frac{l_D}{h_D}$), such as $\lambda = 1$ in the work of Andersson *et al.* [113], $\lambda = 4$ in the work of Pettersson *et al.* [114], or even $\lambda = 174$ as found by Kobatake *et al.* in [115]. Also theoretical studies focused on the role of the aspect ratio λ.

One of the first was by Shinkarenko *et al.* [116], who identifies in the aspect ratio one of the most decisive design parameters whether the texture brings a real improvement to the tribological performance or not. Further results about the optimal aspect ratio are extensively reported in the work of Dobrica and Fillon [46], where the two authors identify the geometrical limitations in the shape of dimple which allow to safely apply the Reynolds equation instead of the Navier-Stokes equations.

With the aim to investigate the relevance of the aspect ratio λ, and more generally the optimal dimple shape, the group at IAM-CMS presented two works focused on the study of individual aspects of the dimple shape in order to systematically determine the influence of each design variable. The details about the experimental set-up used in both above-cited works are given in section 2.1. In the first one, presented by Braun *et al.* [19], the influence of viscosity on the optimal dimple diameter is found by analyzing the optimal dimple diameter at different temperatures with a pin-on-disc set-up. In this way, it was possible to confirm that also the optimal Stribeck curve (*i.e.* the one obtained with the texture which minimizes the friction coefficient) scales with the Hersey number and that the optimal dimple diameter strongly depends on the viscosity value. These experimental evidences laid the basis for part of the numerical work presented in section 4.4. In the second work at IAM-CMS, Schneider *et al.* analyzed the role of the aspect ratio and the texture density, concluding that the optimal values of these two parameters are $\lambda = 10$ and $\rho_{txt} = 10\%$ respectively [20]. This fundings agree with most of the values provided in literature through the research based on similar experimental set-ups [117–119]. The corresponding numerical analysis of this second work of the IAM-CMS group is also performed in the present work in appendix B. From this point of view, this chapter belongs to the branch of tribological research which analyzes the surface textures through a numerical approach. In this context, extensive parametric analyses were carried out by Fowell *et al.* in [120] and by Rahmani *et al.* in [121]. Among the main outcomes, these works pointed out the greater influence that the texture depth has in comparison to its diameter. Moreover, a linear dependence is found between the optimal dimple depth and the reference gap height of the contact geometry [50, 120].

Other strategies, which are pursued in order to find the optimal texture parameters, concern the use of an optimization algorithm based, for example, on non-linear programming routines, [122, 123], or on genetic algorithms [124, 125], or eventually on a combination of both approaches [126]. In the present work we use a parametric approach to find the optimal value, since, in contrast to the majority of literature, the aim is not simply to find an optimal design but also to understand how it scales with respect to the operating parameters, such as velocity, viscosity, gap height and load. Furthermore, this work bases its numerical analysis on the high-resolution representation of the flow in a real set-up geometry through the 2D Reynolds equation with mass conservation cavitation, whilst, to the author's knowledge, all previous optimization studies focus either on 1D cases [120, 121, 124] or on a smaller parameter space [50, 123, 125].

Among the other investigated design parameters, the study of the texture density ρ_{txt} is presented in relation to the one carried out in the work of Schneider *et al.* in [20]. The comparison between numerical and experimental results will be run in view of other literature results, such as those by Wang *et al.* who pointed out how the consideration of contact mechanics results to be crucial for the correct numerical prediction of the role of the texture density [127].

The second experimental campaign, whose results are directly compared to the simulations carried out in this chapter, concerns the influence of partial texturing and, in particular, how the presence of partial textures in different regions of the pin surface impact on the tribological performance of the set-up. In one of the first numerical studies about this topic, Shinkarenko *et al.* underlined how partial texturing can result to be more effective than textures, which cover the whole contact area. Further studies have been presented both experimentally and numerically, by the group of Fillon [48, 49, 128–130] for thrust and plain bearings, by Vlădescu *et al.* [131] for Piston-Liner Contacts and also in the already cited works by Brizmer *et al.* [132] and Guzek *et al.* [122]. In particular, it was found that the textured region generally increases the load carrying capacity, if it is placed in a part of the domain where the gap converges. Moreover, Cupillard *et al.* also identify the region where the gap height is minimum, as the part of the domain where texturing is most convenient, in the case of gap height distributions which are comparable to the one of a pin-on-disc tribometer [133]. From the numerical point of view, this thesis investigates the role of partial texturing in the framework of a sensitivity analysis, *i.e.* by probing the main pin surface with a single dimple in order to compute a sensitivity map which can tell where and how the texture can bring an actual benefit to the overall tribological performance.

The third experimental campaign which is analyzed in this chapter deals with the disposition of the texture elements on the surface and its strong dependence on the texture density. For what concerns the texture disposition, the first studies about it are reported in the works of Yu *et al.* [51, 134], where the authors arranged the texture elements (in the shape of spherical dimple) in different rows and subsequently added a shift to every consecutive row in order to switch from a square pattern to a pseudo-hexagonal one. As a result, it was found that a an optimal angle exists, which maximizes the load carrying capacity and it almost corresponds to a pseudo-hexagonal disposition. Inspired by this findings, Schneider *et al.* analyzed the effects of the texture disposition in a similar fashion from the experimental point of view [20], finding that the pseudo-hexagonal pattern shows the lowest friction coefficient. However, the cited analysis of Yu *et al.* was carried out numerically without considering cavitation and omitting some information about the mesh and other simulation parameters. For this reason, this chapter presents the numerical investigation carried out with the mass conserving implementation of the Reynolds equation presented in section 2.2.2 which was run with the same parameters used in the experiments of Schneider *et al.* in [20].

As conclusion to the introductory part of this chapter, the key point of the numerical analysis in this chapter is the use of a Reynolds solver with mass-conserving cavitation applied on a high-resolution representation of a real pin-on-disc set-up, with the aim to study not only the optimal texture design but also the influence of operating conditions on it. Part of this analysis is also presented in one of the author's publications [CFM$^+$18].

4.1 Numerical representation of the textured pin

As first step, the numerical representation of the textured pin is addressed, with the double aim to show the details of the mesh generation in presence of a texture elements and to elucidate the possible different strategies to apply the Reynolds solver with mass-conserving cavitation for the study of the textured pin.

4.1.1 Texture definition

This subsection introduces the definition of the textures applied on the pin surface which was firstly presented in chapter 2. In particular, the pin geometries considered in this chapter are the so-defined geometries A and B which are introduced in figure 2.3. Among the numerous possible ways to realize textured surfaces, we here focus on non-communicating textures, namely spherical dimples such as those used in the experimental campaigns which were represented in section 2.1 ([19, 20],[CFM$^+$18, GMB$^+$15]). Spherical dimples are one of the most used kind of texture because of their proved effectiveness and because of the fact that they can be described with only two parameters, namely the diameter D and the depth h_D [16, 17], hence reducing the parameter space in the optimization process. Figure 4.1 shows the textured pin surface as considered in most of the simulations carried out in the following analysis. In the figure, the texture is realized through the disposition of spherical dimples according to a pseudo-hexagonal arrangement. In the represented case

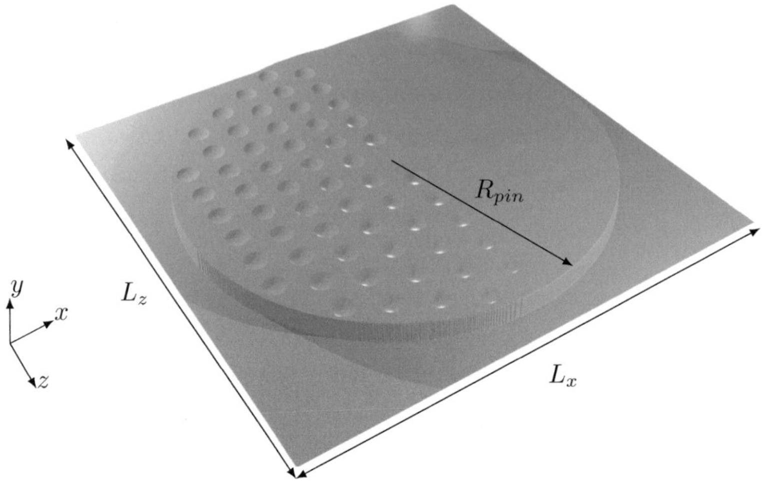

Figure 4.1: Graphical representation of Pin A with texture. The untextured surface distribution is obtained from the measurement of a real specimen used in the experiments by Braun *et al.* in [19], while the texture is generated numerically. The partial front texture here portrayed corresponds to a dimple diameter $D = 400$ μm with a depth $h_D = 40$ μm. The overall texture density is $\rho_{txt} = 10\%$. No texture elements are applied in the center in order to make the reference gap height independent from the texture depth. The domain size is $L_x = L_z = 9$ mm. Adapted from [CFM$^+$18].

only the first half is of the pin surface is textured in order to show better the contrast between the textured and the untextured part. The possible dimple arrangements can be generalized with the aid of the schematic representation shown in figure 4.2. According to this definition of the texture arrangement, every row of dimples is subject to a shift s_z which allows to range from a quadratic pattern ($s_z = 0$) to a pseudo hexagonal pattern ($s_z = 0.5l_z$). A similar definition is used in the experimental work of Yu *et al.* [51] and of Schneider *et al.* [20], where the angle α, as shown in figure 4.2, is used to identify the type of pattern. It is important to notice that the distances between two adjacent dimples are kept equal to each other in all simulations, namely $l_x = l_z$. This allows to keep the texture density ρ_{txt} independent from the considered arrangement. For this reason, the arrangement obtained for $s_z = 0.5l_z$ assumes

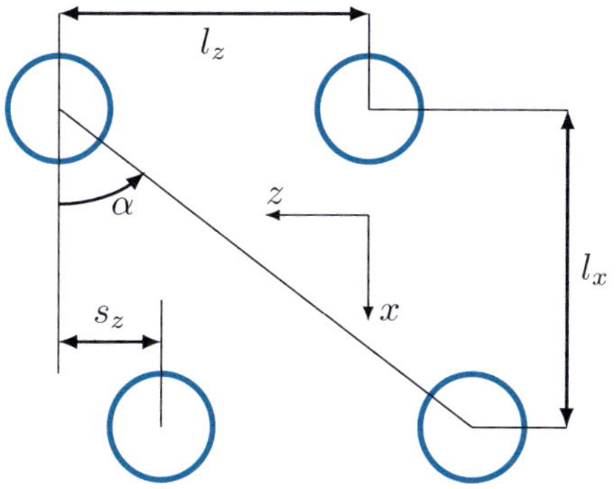

Figure 4.2: Schematic representation of the texture arrangement. The upper wall slides with constant velocity U in the x-direction. The square pattern is obtained for $s_z = 0$, while the pseudo hexagonal occurs for $s_z = 0.5l_z$. The horizontal and the vertical spacing are always equal to each other $l_x = l_z$. Adapted from [CFM+18].

the shape of a slightly distorted hexagon, whose bases measure l_z while the other four sides measure $\frac{\sqrt{5}}{2}l_z$. This particular arrangement definition is referred to as pseudo-hexagonal in the following discussion.

In summary, the considered texture parameters are four: dimple depth h_D, dimple diameter D, dimple spacing in the streamwise and spanwise direction l_x and l_z and the shift of the texture rows s_z. Other design parameters, such as the texture density ρ_{txt} are indirectly defined as result of the combination of D, l_x and l_z.

For what concerns the numerical solution of the cases analyzed in this chapter, the Reynolds solver with mass conserving cavitation (see section 2.2.2) was employed in all the presented simulations. This leads to the particular challenge of dealing with a very broad range of scales, ranging form the pin size ($R_{pin} = 4$ mm) to the smallest texture diameter $D = 40$ μm. The efficient implementation of the mass conserving cavitation algorithm has allowed to mesh the 2D pin domain shown in figure 4.1 with up to 2049 cells per side, reaching a maximum resolution of 4.39 μm in both directions while keeping the solution time in the order of few minutes (in case of a single core simulation).

number of cells in each direction	normal force $[N]$	tangential force $[N]$
65	1178.379057	5.029480
129	1188.524679	5.037692
257	1193.755952	5.045461
513	1196.401560	5.049900
1025	1197.470182	5.051418
2049	1198.008821	5.052289
4097	1198.288155	5.052802

Table 4.1: Convergence of the Reynolds solver with respect to the normal and tangential force. The shown values correspond to the textured pin geometry A with a dimple diameter $D = 40\mu m$ which represents one of the most strict cases because of the relatively small size of the dimples. Due to the negligible difference between the last two cases, a resolution of 2049 cells per side was chosen as standard for the simulations in this chapter. Adapted from [CFM+18].

The spatial convergence analysis for the textured pin A is reported in table 4.1. Similar convergence rates are obtained also with the second type of pin, namely the geometry B. Further details about the computational performance are reported in section 2.2.2.

The numerical solution through the Reynolds solver presented in section 2.2.2 delivers two fields, namely the pressure distribution $p(x,z)$ and the cavity fraction $\theta(x,z)$. Figure 4.3 and 4.4 show, respectively, the pressure and the cavity fraction distribution for the textured case whose mesh is shown in figure 4.1. These distributions correspond to a typical solution scenario in which the first half of the pin experiences a strong pressure increase while cavitation dominates the second half of the pin domain. This is due to the particular profile of the pin A (see figure 2.3) which presents a certain curvature, converging in the front and diverging in the rear. The cavity fraction shown in figure 4.4 is directly related to the density of the lubricant according to equation 2.22 and shows how the density decreases in the cavitation region in order to preserve the mass conservation. It is important to notice that the domain boundaries

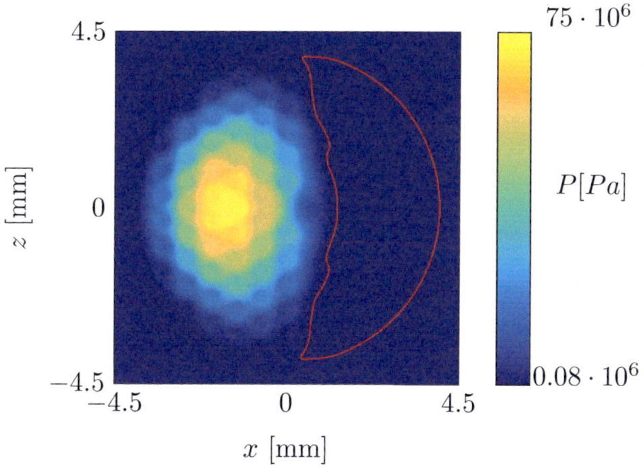

Figure 4.3: Pressure distribution over the textured pin surface shown in figure 4.1 with the following texture parameters $D = 400$ μm, $h_D = 40$ μm and $\rho_{txt} = 10\%$. The red line delimits the cavitation region. The chosen operating parameters are the following: the upper wall moves from left to right with velocity $U = 1$ m/s, the dynamic viscosity is $\mu = 0.1871$ Pa·s and the gap height is $H = 1$ μm. Adapted from [CFM⁺18].

are far enough away from the pin edge, so that both the pressure and the cavity fraction distribution are not affected by the imposed boundary conditions. This is also due to the big ratio between the height of the pin edge and the average gap height on the pin surface. The importance of a relatively big gap height of the inlet in comparison to the gap height in the rest of the domain was pointed out also by Cupillard *et al.* in [133] as a geometrical condition in order to make the solution independent from the boundary conditions.

4.1.2 Setting of the numerical parametric study

In order to ease the comparison, numerical simulations should mimic as well as possible the same operating conditions which characterize the experimental measurements with the pin-on-disc set-up. In the experiments, the tribological tests are run under the condition of a constant normal force F_N and the gap height, which is unknown, is determined by the resulting equilibrium. On

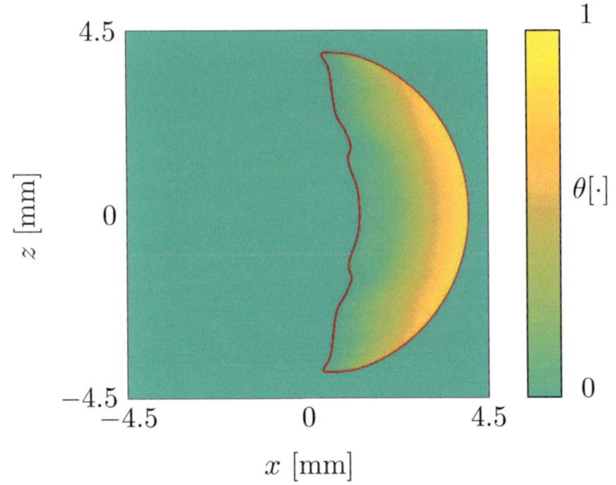

Figure 4.4: Pressure (a) and cavity fraction (b) distribution over the textured pin, $D = 400$ $[\mu m]$, $Depth = 40$ $[\mu m]$, $\rho_{txt} = 10\%$. The upper wall slides from left to right with velocity $U = 1m/s$, viscosity $\mu = 0.1871 Pas$ and the gap height $H = 1\mu m$.

the other side, the Reynolds equation requires the complete gap height distribution as input, providing the normal force as result of the integration of the pressure distribution. In order to cope with the intrinsic difference of the two approaches, an iterative algorithm can be used, so that the simulations with the Reynolds equation converge to the a solution which fulfills a prescribed value of normal force by adapting the gap height at each iteration. The introduction of such an iterative algorithm can result computationally expensive if applied to parametric studies [16]. For this reason, simulations with prescribed normal force will be considered in this work only for a certain part of the following numerical analysis. In particular, simulations with prescribed normal force will be considered in section 4.4 where the direct comparison with experimental data strictly requires such an approach. The root-finding algorithm chosen in this this work is the secant algorithm, which represents a good compromise between stability and computational speed [135].

As mentioned in section 2.1, an important aspect of the numerical representation of the pin-on-disc tribometer is the self-aligning pin holder which allows the pin to pitch in order to prevent misalignment errors during the mounting and assure a flat-on-flat contact [19]. Due to the high asymmetry of the pressure distribution over the pin, a pitching-moment can occur which can consequently make the pin pitch of a certain angle and, hence, changing the overall gap height distribution drastically. This pitching movement is further complicated by the adhesion mechanism between the surface of the pin holder and the lower support of the tribometer. This makes the problem of a numerical representation of the whole pin-on-disc device more complicated and computationally expensive, because of further iteration loop which is needed in order to find the resulting pitch angle which change the inclination of the pin until the pitch-moment becomes zero. Therefore, the consideration of both iterative loops concerning the gap height and the pitch angle would make an extensive parametric study on textures unfeasible. This is evident, for example, in the recent work of Gropper *et al.* [125], where the consideration of both tilting angles of a textured pad bearing inevitably leads to the restriction of the number of considered texture elements and other investigated texture parameters.

4.2 Effects of the number of dimensions of the Reynolds equation

In this section the relevance of the numerical approach based on the 2D Reynolds equation is assessed. The aim is to investigate the differences between the 1D and the 2D representation of a textured surface, in order to understand how much the quality of the numerical simulations can take advantage from a 2D approach despite the higher computational costs. Moreover, particular attention is also paid to the influence of the macro geometry of the pin surface on the pressure distribution due to the presence of the texture. This is done for the first time with a realistic pin-on-disc set-up, while previous literature works, which carry out such systematic investigations, mainly

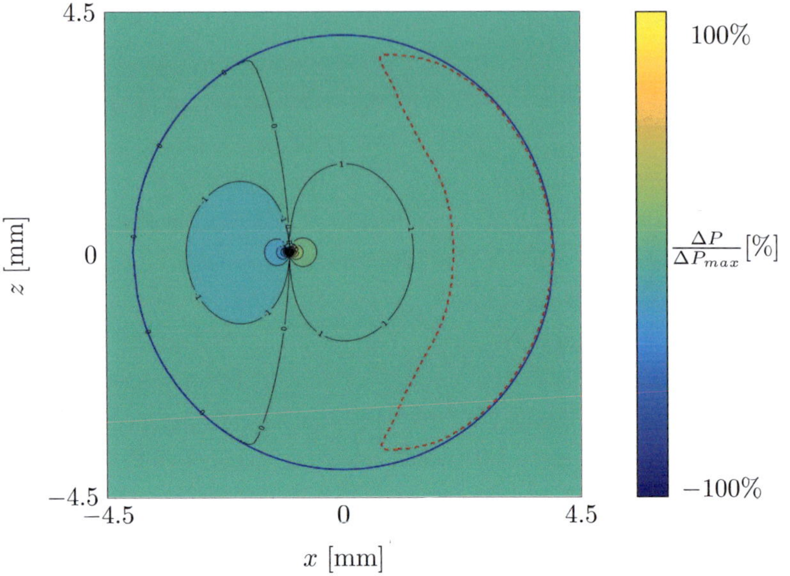

Figure 4.5: Impact of the presence of a dimple on the net normalized pressure distribution over the pin surface (marked within the blue line). The normalized net pressure distribution is computed as $\frac{\Delta p}{\Delta p_{max}}$ where $\Delta p = p_{txt} - p_{untxt}$. The region in which cavitation occurs is marked with the red dotted line. The same operating parameters were used as in figure 4.6 for the case with $x_c = -\frac{3}{4}R_{pin}$. Adapted from [CFM+18].

describe either 1D slider bearings [21, 120, 121, 136], 1D or 2D parallel thrust bearings [124, 137–139], 2D seals [140] or 2D journal bearing [141, 142].

Figure 4.5 shows the pressure perturbation over the pin surface due to the presence of a single dimple. The pressure distribution shown in figure 4.5 is computed through the ratio $\frac{\Delta p}{\Delta p_{max}}$ where $\Delta p = p_{txt} - p_{untxt}$. It can be noted, that pressure perturbation due to the dimple decays very fast and gets smaller than 1% within 10 diameters from the center of the dimple.

Contrariwise, when the dimple is placed on a 1D representation of the pin, the influence of its pressure perturbation is propagated to a much farther distance. This can clearly be seen in figure 4.6, where the net 1D pressure perturbation due a single dimple is directly compared to the centerline pressure distribution extracted from figure 4.5. In particular, the pressure perturbation in the 1D case also has a different decay rate than in a 2D simulation and this could deeply influence the computation of the resulting load carrying capacity, hence leading to a false estimation of the latter. In both 1D and 2D cases, the net pressure distribution becomes almost zero when it reaches the cavitation region.

The difference between the two cases may be attributed to the fact that a 1D dimple is not really representative for a spherical dimple but it rather represents a cylindrical channel with infinite extension in the spanwise direction. Moreover, the pressure gradient in the z-direction contributes to the attenuation of the pressure perturbation due to the presence of the dimple.

This analysis on the effects of the number of the dimension shows that the application of a 1D approach may lead to wrong estimations of the impact of textures. This evidence has, surprisingly, aroused little attention in the research community, since, most of the works, which compared simulations carried out with a different number of spatial dimensions, focused rather on the differences between the 2D and the 3D Navier-Stokes equations [45, 143, 144], although the Reynolds equation is by far the most used approach in literature [16].

4.3 Sensitivity analysis

This section focuses on the influence of the position of the texture elements on the pressure distribution over the pin. The goal is to deduce important insights about the interaction between the texture elements and the macroscopic geometry of the pin-on-disc set-up. This investigation can be seen as a sensitivity analysis, in which the pressure perturbations due to the presence of texture elements are systematically studied also under different operating conditions, in order to generalize the findings as much as possible.

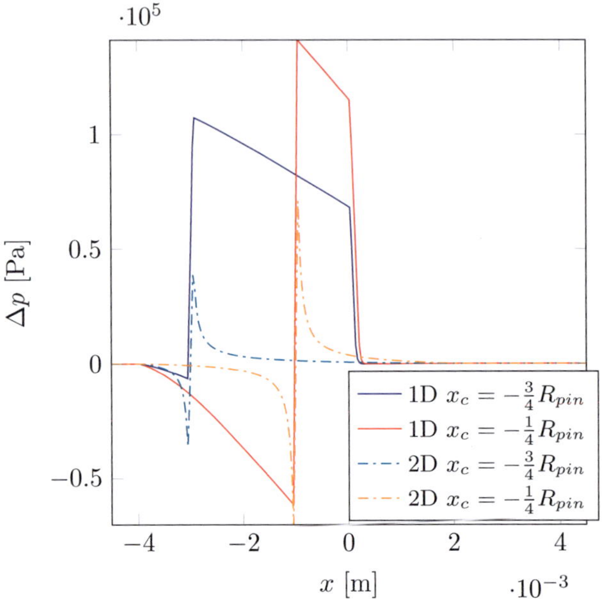

Figure 4.6: Comparison of the pressure distribution in the centerline of the pin between 1D and 2D simulations. The figure shows the difference between the pressure distributions with the textured surface and the untextured case $\Delta p = p_{txt} - p_{untxt}$. Two different dimple positions on the first half of the pin surface are shown, in the second half cavitation occurs. Both simulations are carried out with the following parameters: $H = 10$ μm, $\mu = 0.18$ Pa, $U = 0.1$ m/s. Adapted from [CFM+18].

In the first part of the sensitivity analysis, a group of texture elements is studied at once, while the second part focuses rather on the effects generated by single dimples. In particular, the effects of texture on the half pin surface will be addressed first, since the numerical investigation can be easily placed side by side with the experimental activity carried out at the IAM at KIT [19, 20],[CFM+18].

Experimental tests with single dimple are unfeasible due to the risible signal-to-noise ratio, therefore, the role of the position of the textured area has to be investigated by means of clusters of dimples on the surface. In this regard, an experimental analysis is presented out in [CFM+18], by focusing on two

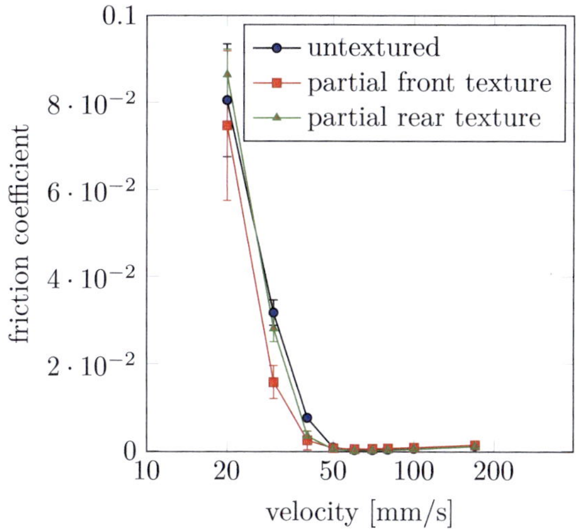

Figure 4.7: Experimental Stribeck curve for different extensions of the textured area. Three different cases are shown, namely a partial texturing on either the front or the rear part of the pin and the untextured pin. In both the textured cases the texture consists of a spherical dimple in a pseudo-hexagonal arrangement with the same design parameters as in the numerical simulations shown in figure 4.8: $D = 60$ μm, $h_D = 4.5$ μm, $\rho_{txt} = 10\%$. The error bars represent the standard deviation. Adapted from [CFM$^+$18].

different extensions of the textured area, namely only on the leading or on the trailing half of the pin. Figure 4.7 shows the experimental Stribeck curve obtained for the two mentioned cases and the reference untextured one. The texture consists of spherical dimples arranged in a pseudo-hexagonal pattern with the following characteristics: $D = 60$ μm, $h_D = 4.5$ μm, $\rho_{txt} = 10\%$. As one can see, the texturing of the leading half leads to a friction reduction in comparison to the other two curves. On the other side, texturing the trailing part of the pin results in a smaller friction reduction and may actually lead to an increase of the friction coefficient at small velocities. These results are in agreement with previously-cited studies in literature [49, 130, 131] about partial texturing. The description of the set-up configuration used to obtained the curves in figure 4.7 are shown in detail in section 2.1. In particular, it has to be

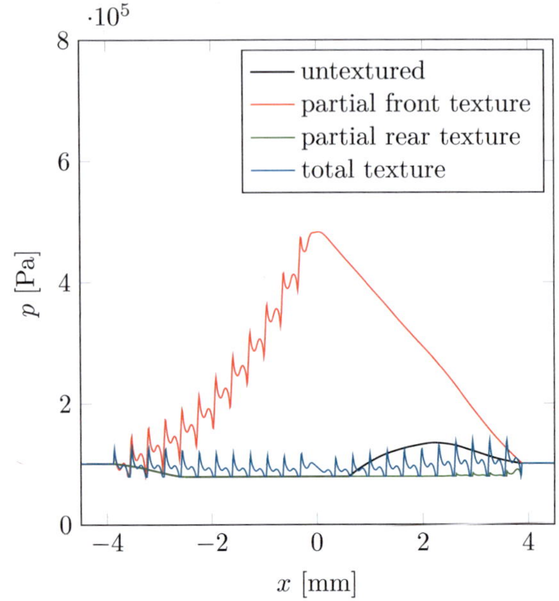

Figure 4.8: Pressure distribution in the centerline of the pin for different extensions of the textured area. This simulations are carried out with the pin geometry B in accordance to the experiment carried out in [CFM+18] and presented in section 2.1. The operating parameters used for the simulations in the figure are: $H = 10$ μm, $U = 0.1$ m/s, $\mu = 0.18$ Pa·s. Adapted from [CFM+18].

noted that the measurements focused mainly on the mixed lubrication region, because the very low signal-to-noise ratio does not allow to appreciate the friction differences in the hydrodynamic regime with a sufficient level of certainty.

The numerical counter part of the results in figure 4.7 is shown in figure 4.8, where the pressure distribution over the centerline of the pin is portrayed for four different cases: front and rear texture, total texture and, finally, the untextured case as reference. Only the partial texture in the front part of the pin surface leads to a net increase of the load carrying capacity. In this case, the rear texturing can even lead to a decrease of the performance in comparison to the untextured case, since the corresponding pressure distribution is clearly lower and, hence, with a smaller resulting load carrying capacity. Since a higher load carrying capacity can typically be associated with a smaller friction, these results can be related to the experimental ones in figure 4.7, where the partial front texturing showed a better impact on the reduction of the friction coefficient than the rear one. Both experiments and the numerical simulations were obtained with the pin geometry B (see figure 2.3), whose very flat surface could resemble the one of other applications, such as parallel bearings. This similarity allows to run comparisons with previous literature results, which also found that texturing the front part of such a geometry results more effective than full or rear texturing [49, 130, 131].

Having assessed the validity of the numerical approach for the study of the sensitivity analysis through the comparison with experimental results concerning textures with multiple dimples, we move the focus to the sensitivity analysis by means of single texture elements. In particular, the goal is to study how the perturbations of the load carrying capacity due to the presence of the dimple are influenced by the position of the dimple itself on the surface and by the operating parameters. For this reason, several simulations are carried out for different values of the operating parameters (gap height, viscosity, applied load and velocity of the upper wall) and of the texture design parameters, such as the diameter and the depth. The operating parameters can be represented in a more generic way by recalling the definition of the Sommerfeld number in equation 2.8:

$$\mathcal{S} = \frac{\mu U L_x}{H^2 W}. \tag{4.1}$$

In the following analysis, the reference domain length L_x does not vary and the reference applied load W is the one from the reference untextured case, since the presence of a single dimple affects the overall load carrying capacity in a negligible way.

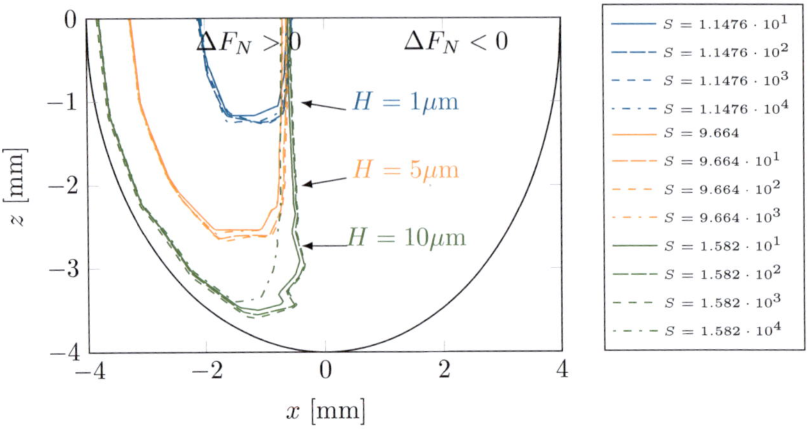

Figure 4.9: Zero-isolines of the net normal force $\Delta F_N = F_{N_{txt}} - F_{N_{untxt}}$ obtained through the variation of the operating parameters. The portrayed isolines are a function of the position of the center of the dimple on the surface of the pin. The variation of the operating parameters is summarized by the Sommerfeld number in the legend. The values of viscosity, velocity and gap height are varied in the following ranges: $0.0187 < \mu < 1.871$ Pa·s and $0.01 < U < 1$ m/s, $1 < H < 10$ μm. Due to the symmetry of the geometry, only the lower half of the pin is portrayed here and the flow comes from left to right. The simulations are carried out with prescribed gap height and the normal load is an output. Adapted from [CFM+18].

At first, we focus only on the operating parameters by using a probing dimple with a fixed size of $D = 100$ μm and $h_D = 10$ μm. This size can be considered small enough to introduce small perturbations which are, at the same time, big enough in order to have a noticeable and clear effect. The results of this first part are presented in figure 4.9, where the zero-isolines of the net normal force are shown as a function of the position of the center of the probing dimple on the surface. Twelve different isolines are represented in three groups, each group represents a different gap height ranging from 1 to 10μm. The four lines in each groups represent, in turn, a different value of the Sommerfeld number obtained through the different combination of viscosity and velocity.

It is important to notice that the contour area with positive normal force concerns always the first half of the pin surface. This hints to the fact that in the presence of a total texture, only the front part can actually increment the load carrying capacity, while the rear part could have a detrimental effect. This is also due to the fact that the dimples have almost no effect in the cavitation zone, because their influence on the pressure distribution is not big enough to influence the cavitated regime in most of the rear part of the pin. These conclusions are again in agreement with the above-mentioned literature results [49, 130, 131] and also with the experimental and numerical analysis on the partial texturing which was previously assessed in this section. Furthermore, as pointed out by Cupillard *et al.*, in case of geometries with a curvature similar to that of pin A, dimples are deemed to be more adequate if placed close to the center of the region where the pressure distribution reaches its maximum [133].

In figure 4.9, isolines with the same gap height but different values of the Sommerfeld number (*i.e.* with different values of viscosity and velocity) almost collide. On the other hand, isolines with a different gap height are much farther away from each other. This means that the gap height is a much more sensitive parameter than velocity and viscosity. As a matter of fact, small variations of the gap height have a much bigger impact on the distribution of the area corresponding to an actual benefit introduced by the probing dimple. In summary, if the Sommerfeld number changes because viscosity or velocity vary, the region, in which a dimple would increase the load carrying capacity, remains almost unchanged.

If the gap height is very small the positive net force region shrinks considerably and, for the considered cases, may approach to zero if the $H < 0.1$ μm. In this regard, the work of Murthy *et al.* presented a similar consideration by analyzing slider bearings [136]. Indeed, the smaller the gap height in the pin geometry, the higher becomes the slope of the gap height due to the curvature of the pin surface and this effect is known to reduce the area where dimples bring actual benefits in slider bearings which are highly slanted.

The second part of the sensitivity analysis with a single probing dimple concerns the role of the dimple design parameters, *i.e.* diameter D and depth h_D (as well as the aspect ratio $\lambda = \frac{D}{h_D}$). The impact of the dimple diameter and aspect ratio of the probing dimple are shown in figure 4.10. In particular, figure 4.10a represents the sensitivity analysis carried out with different dimple diameters and depths but keeping the aspect ratio constant. In this case, for each group of isolines with a different gap height, the curves do not collide, meaning either the diameter or the depth (or a combined effect of both of them) changes the influence area of the probing dimple. In order to cast light on this aspect, figure 4.10b shows the simulation results where the dimple depth is kept constant to $h_D = 10\ \mu\text{m}$ and only the diameter is varied. As one can see, the three isolines with the same color (corresponding to different diameters for every investigated gap height) collide. This means that the extension of the region where dimples work is not affected by the diameter of the dimple, therefore, the dimple depth is responsible for the deviations seen in figure 4.10a. This suggests that the dimple depth has a much higher impact on the properties of the texturing than the dimple diameter.

4.4 Scaling of the optimal dimple depth

This section introduces the two main parametric studies, which were carried out in order to analyze the scaling of the optimal dimple shape, both in terms of maximal load carrying capacity and minimal friction coefficient, with respect to the operating conditions. The two parametric studies are based on the two possible strategies which can be employed in the numerical simulation of pin-on-disc tribometers, namely the carrying out of simulations with prescribed gap height H or with constant normal force F_N. The first simulation strategy has the great advantage of being fast and easier to set than the second one, hence enabling us to focus on a richer parameter space. On the other side, the second strategy mimics more realistically the experimental set-up, where

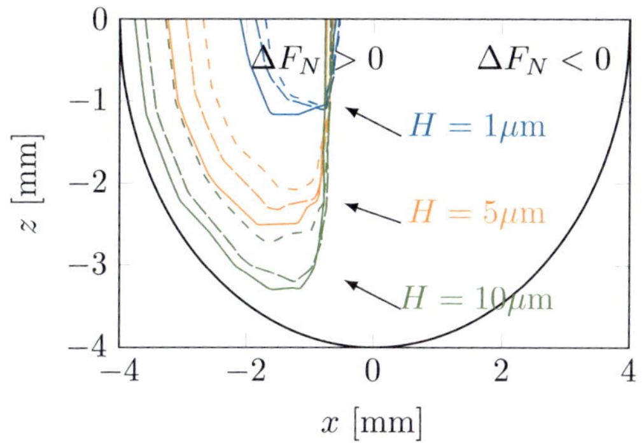

(a) Sensitivity analysis with constant dimple aspect ratio $\lambda = 10$ but different dimple diameters and depths (——) $D = 100$ μm and $h_D = 10$ μm, (−−·) $D = 200$ μm and $h_D = 20$ μm, (- - -) $D = 400$ μm and $h_D = 40$ μm.

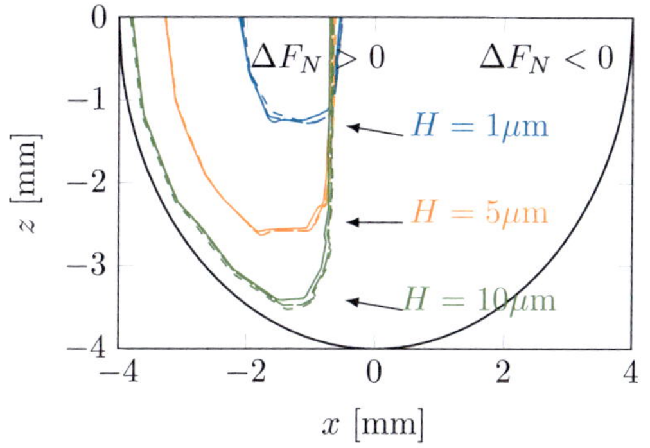

(b) Sensitivity analysis with constant dimple depth $Depth = 10$ μm but different dimple diameters (——) $D = 100$ μm, (−−·) $D = 200$ μm, (- - -) $D = 400$ μm.

Figure 4.10: Zero-isolines of the net normal force $\Delta F_N = F_{N_{txt}} - F_{N_{untxt}}$ as function of the position of the dimple on the pin surface. The two subfigures present the results concerning the sensitivity analysis with different texture parameters, such as dimple diameter D and dimple depth h_D. The isolines are divided in three groups with different colors, each color corresponds to a specific gap height in the range $1 < H < 10$ μm. Viscosity and velocity are kept constant in this plots: $\mu = 0.187$ Pa·s and $U = 0.1$ m/s. Adapted from [CFM+18].

a constant known load is applied on the upper disc and the gap height is the result of the force equilibrium.

The motivation to numerically investigate the scaling of the optimal shape comes from the experimental evidences about the relationship between the optimal dimple diameter and the viscosity presented by Braun *et al.* in [19]. Figure 4.11 shows the experimental Stribeck curves measured by Braun *et al.* at two different temperatures (*i.e.* with two different values of viscosity). For both temperatures, the untextured case is compared to the one whose texture diameter entails the lowest friction coefficient. The aspect ratio is $\lambda = 10$, and is a constant parameter during the experimental campaign. Therefore, also two optimal depths, namely $h_{D_{opt}} = 4$ μm at $T = 100$ °C with $\mu = 0.00308$ Pa·s and $h_{D_{opt}} = 20$ μm at $T = 50$ °C with $\mu = 0.1117$ Pa·s, correspond to the two optimal diameters.

This experimental results will be further analyzed in comparison to the numerical simulation with constant normal force in the following subsection (4.4.2).

4.4.1 Analysis of the optimal dimple depth

This subsection is based on the numerical analysis of the tribometer set-up with prescribed gap height. This first parametric study focuses on the influence of different dimple diameters and depths on the load carrying capacity and on the friction coefficient. As presented in [69] and [145], the computation of these two quantities for a flow with cavitation is based on the following definitions:

$$F_N = \int_A (p - p_{amb})\, dA, \tag{4.2}$$

$$F_T = \int_A \left[(1-\theta)\,\mu \frac{U}{h} + (1-\theta)\,\frac{h}{2}\frac{\partial p}{\partial x} \right] dA, \tag{4.3}$$

$$C_f = \frac{F_T}{F_N} \tag{4.4}$$

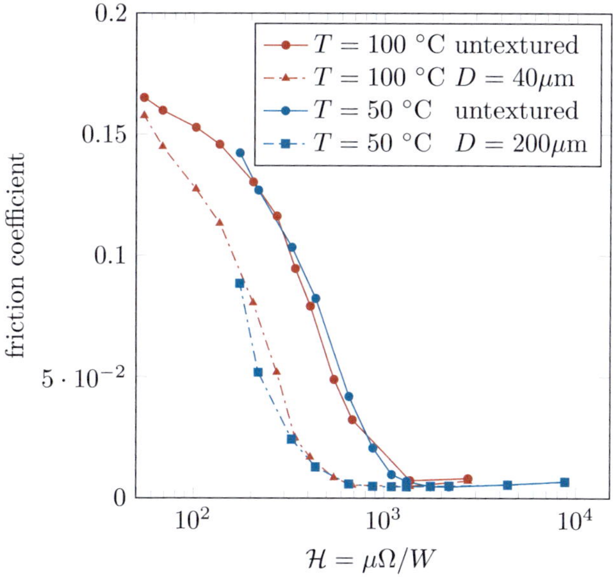

Figure 4.11: Stribeck curve of the experimental investigation of the optimal dimple diameter performed by Braun *et al.* [19]. Two groups of curves are portrayed, each one at a different test temperature, hence two different viscosity values. For each of the two groups, both the untextured case as well as the case with optimal texture diameter are represented. The dimples were obtained with constant aspect ratio $\lambda = 10$, meaning that the optimal depth values are $h_{D_{opt}}|_{T=100^\circ} = 4$ μm and $h_{D_{opt}}|_{T=50^\circ} = 20$ μm.

where p_{amb} is the ambient pressure which is considered to act homogeneously on both sides of the rotating disc.

For each point in the diameter-depth space different operating conditions are tested by varying viscosity, velocity and gap height in a broad range which is representative for running conditions in typical tribological applications. The investigated range for the viscosity and velocity values is resumed in table 4.2 while the gap height H was varied between 0.5 and 20 μm.

As shown in section 4.3, only the first part of the pin surface experiences a real benefit if textured. For this reason, only a partial texture in the front part

U [m/s] / μ [Pa·s]	0.01	0.05	0.1	0.5	1	5	10
0.0187			■		●		♦
0.0935					●		
0.1871	■	*	■	×	●	⬟	♦
0.9357					●		
1.8714	■		■		●		♦

Table 4.2: Velocity and viscosity range considered in the parametric study with prescribed gap height. The values in the table correspond to those portrayed in figure 4.13. As in the experiments, the texture density is kept constant at $\rho_{txt} = 10\%$, while the gap height ranges between 0.5 and $20\mu m$. The overall range of Sommerfeld number based on the gap height H is $172 < S < 14.3 \cdot 10^3$.

of the surface is considered in this part of the present work. The texture density is kept constant at $\rho_{txt} = 10\%$ as in the experiments from Braun *et al.* [19]. The dimple diameter ranges from 40 to 400 μm while the depth varies between 0 (untextured case) and 200 μm. Since the dimple has a spherical shape, the minimum aspect ratio cannot exceed $\lambda = 2$. Following the preliminary results presented in section 2.3.3, the applicability of the Reynolds equation may become arguable for values of the aspect ratio which are below $\lambda < 10$. The implications of this condition will be addressed further while presenting the results in the following part of this subsection. Moreover, the Reynolds number never exceeds the value $Re = 0.01$ for every combination of the operating parameters.

The diameter-depth parameter space is now analyzed for each possible combination of the operating parameters, namely U, μ and H. As a result, figure 4.12 shows the normal force as a function of the dimple depth and diameter for a certain combination of the operating parameters. It can be seen, at first, that the distribution of the normal force is more affected by variations of the depth than of the diameter. In particular, the shape of the normal force distribution clearly presents a maximum value which is aligned with a certain

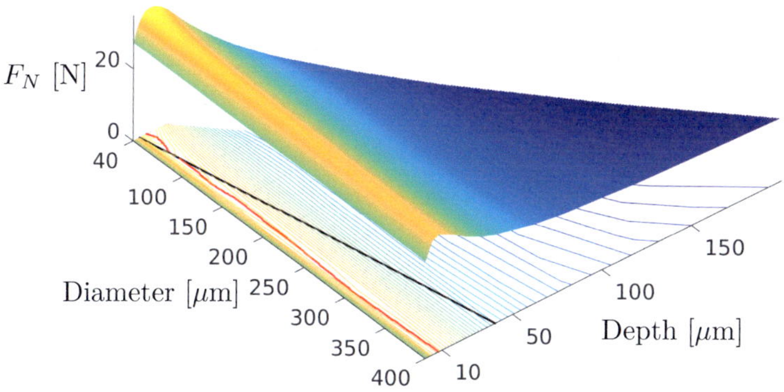

Figure 4.12: Distribution and contour lines of the normal force in the depth-diameter parameter space. The red line marks the position of the optimal depth $h_{D_{opt}}$ which maximizes the normal force F_N. The black line marks the position where dimples have an aspect ratio $\lambda = 10$ which is the one considered in the experiments by Braun *et al.* [19]. Moreover, most of the optimal depth line (red) is above the ratio $\lambda = 10$, hence ensuring the validity of the Reynolds equation used in this analysis. The shown case is computed with the following parameters: $U = 1$ m/s, $\mu = 0.187$ Pa·s, $H = 7\mu$m. Adapted from [CFM$^+$18].

constant depth of the dimple. This means that an optimal dimple depth $h_{D_{opt}}$ can be identified and this value remains insensitive to variations of the diameter. The value of the optimal depth is marked with a red line in the projected contour plot in figure 4.12.

The black line in the figure corresponds to the aspect ration $\lambda = 10$. As one can see, the great part of the optimal depth line lies in the region of the parameter space where the aspect ratio is bigger than 10, meaning that the most relevant part of the depth-diameter parameter space lies in the applicability region of the Reynolds equation, where the condition on the aspect ratio is fulfilled [46].

The absolute value of the normal force has a direct proportionality on velocity and viscosity through the Sommerfeld number. Nonetheless, the shape of the normal force distribution does not depend on the operating conditions and one can always clearly identify a maximum in the depth-diameter space which is aligned with a certain value of the dimple depth and does not almost depend on the diameter. Only by varying the gap height H one can see a shift in the normal force distribution which will simply change the location of the maxima but still keep the same distribution shape, hence allowing to find an optimal dimple depth at a shifted value.

The so-identified optimal depth can be plotted as function of the gap height for different values of viscosity and velocity as shown in figure 4.13. Each point in the figure corresponds to the optimal depth value which is computed with a certain gap height for a specific value of viscosity and velocity whose range is presented in table 4.2. It is interesting to notice that the optimal depth is, with good approximation, a linear function of the gap height, regardless of the analyzed combination of viscosity and velocity. This linear dependence is also in agreement with literature results based on the 1D Reynolds equation, such as those by Fowell *et al.* [120] and ramesh *et al.* [50].

This analysis hints to the fact that geometrical parameters such as the gap height, play a more important role than viscosity and velocity in the definition of the optimal texture shape. However, it has to be noticed that this statement holds only for simulations which are carried out with prescribed gap height. In case of simulations with constant normal force, the variations of the viscosity and of the velocity cause changes in the gap height and, hence, in the definition of the optimal dimple depth $h_{D_{opt}}$.

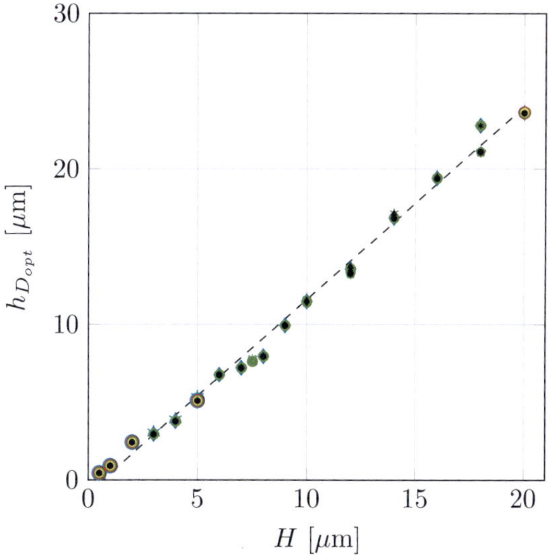

Figure 4.13: Optimal dimple depth as function of the gap height. Each point in this graph represents the optimal depth extrapolated from the depth-diameter parameter space as show in figure 4.12. The plot style of the points corresponds to the respective operating parameter shown in table 4.2. Adapted from [CFM+18].

4.4.2 Scaling between the dimple depth and the Sommerfeld number

In order to analyze the dependence of the optimal texture design on the operating parameters, this section extends the previous work for the simulation approach with constant normal force. This simulation approach is in agreement with the working principle of the experimental set-up and, therefore, the most suitable approach in order to compare the numerical results with the above-cited experimental ones by Braun *et al.* [19].

In the simulations presented in this subsection, the normal force is prescribed and the gap height is iteratively varied through the application of the secant algorithm. The normal force applied on the rotating disc is directly related to the applied normal load through the area of the disc as follows:

$$W = F_N A. \tag{4.5}$$

This relationship can be further developed in order to make the dependence of the applied load W on the operating conditions emerge. It is known [71, 146] that the pressure distribution scales with the following law:

$$p = p_r \frac{\mu U L_x}{H^2} \tag{4.6}$$

where p_r is an arbitrary reference pressure. By combining the normal force definition (equation 4.2) with the definition of applied load (equation 4.5), one can rewrite the above equation as follows:

$$W = W_r \frac{\mu U L_x}{H^2} \tag{4.7}$$

where, similarly to p_r, W_r is an arbitrary dimensionless reference load. Consequently, the variation of the gap height can be expressed as:

$$H = \sqrt{\frac{\mu U L_x W_r}{W}}. \tag{4.8}$$

Equation 4.8 implies that, in case of simulations with prescribed normal force F_N (applied load W), any change in the viscosity and the velocity will cause a variation in the gap height H.

With this simulation settings, the depth-diameter space is investigated again as previously done in the simulations with prescribed gap height (subsection 4.4.1). In this case, since the normal force is prescribed, the search for the optimal depth is carried out by analyzing the distribution of the friction coefficient in the depth-diameter space. Figure 4.14 shows the distribution of the friction coefficient as a function of the depth-diameter space. As happened in

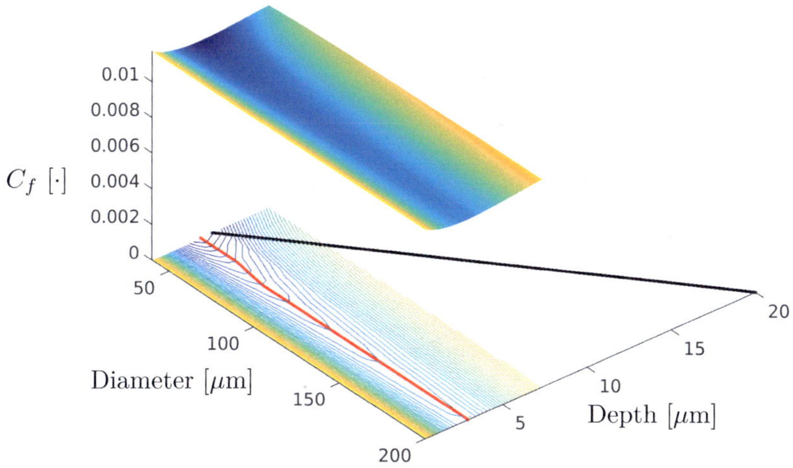

Figure 4.14: Distribution and contour lines of the friction coefficient in the depth-diameter parameter space. The red line marks the position of the optimal depth $h_{D_{opt}}$ which minimizes the friction coefficient C_f. Similarly to figure 4.12, the black line marks the position where dimples have an aspect ratio $\lambda = 10$ which is the one considered in the experiments by Braun *et al.* [19]. The depicted case is computed with the following parameters: $U = 1$ m/s, $\mu = 0.187$ Pa·s, $F_N = 20$ N.

the previous subsection, an optimal depth can be identified which can minimize the friction coefficient. Moreover, the value of the optimal depth does not depend on the diameter with good approximation.

In this case, the value of the optimal depth changes with the viscosity and velocity values. This change is, however, only due to the fact that, since the applied normal force is constant, the gap height changes according to the variation in viscosity and velocity. As a matter of fact, if the optimal depth value

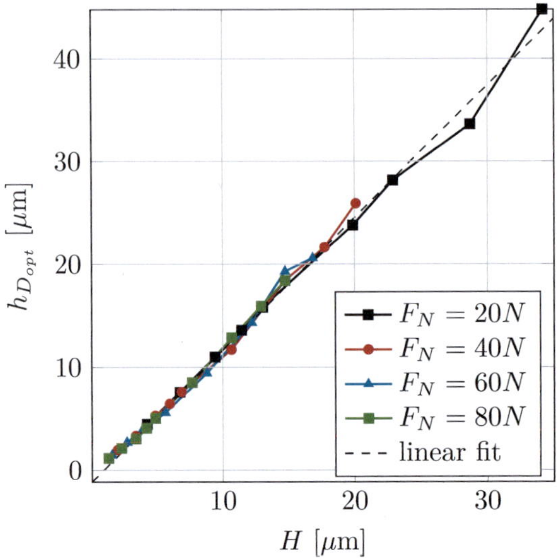

Figure 4.15: Optimal dimple depth as function of the gap height for simulations with constant normal force. The four colored lines represent the optimal depth extrapolated from the depth-diameter parameter space shown in figure 4.14 for the case with constant normal force. Adapted from [CFM+18].

$h_{D_{opt}}$ is plotted as function of the resulting gap height H the same linear relationship can be clearly seen as in the previous subsection. Figure 4.16 shows this relationship for different values of the normal force, which ranges from 20 to 80 N. The optimal gap values are extracted from the depth-diameter space (as shown in figure 4.14) in the same way as done in the previous subsection.

The linear relationship between the optimal depth and the gap height can be explicitly formulated by recalling, at first, a more general description of how the optimal depth depends on the other parameters:

$$h_{D_{opt}} = h_{D_{opt}}(\mu, U, H, D, W). \tag{4.9}$$

This general formulation can be rewritten according to the results shown in this section. Firstly, figures 4.12 and 4.14 show that the optimal depth does not depend on the diameter with good approximation. Therefore, the dependence on the diameter can be neglected in equation 4.9. Secondly, the optimal depth is found to scale linearly with the gap height for both the possible simulation approaches (constant applied force or prescribed gap height). Thanks to these two findings one can rewrite equation 4.9 as follows:

$$h_{D_{opt}} = kH(\mu, U, W) \tag{4.10}$$

where k is a constant which depends only on the shape of the macro-geometry. At this point, the dependence of the gap height on the operating conditions is known from the above-mentioned equation 4.8. Therefore, by substituting equation 4.8 into equation 4.10 one obtains

$$h_{D_{opt}} = k\sqrt{\frac{\mu L_x U W_r}{W}} = K\sqrt{\frac{\mu L_x U}{W}} \tag{4.11}$$

where the reference load W_r has been grouped outside the square root together with k into the a new constant K. The reference length L_x is considered, in this analysis, as a constant because it always refers to the domain length of the pin geometry. It appears clear now that the optimal depth is proportional to the square root of viscosity, velocity and reference length, while it is inversely proportional to the square root of the normal load W.

The dependence of the optimal depth on the square root of the viscosity is confirmed in figure 4.16 where the optimal depth values obtained from the procedure shown in figure 4.14 are plotted against the viscosity. Four curves are shown from the simulations with four different values of normal force. A dashed line representing the square root dependence of the optimal gap height is introduced in the figure in order to show the good agreement between the simulation results and the scaling proposed in equation 4.11.

Moreover, the optimal depth values, which were found in the experimental work of Braun *et al.* [19], are also plotted in figure 4.16. These experimental values are obtained form figure 4.11 by converting the diameter values into

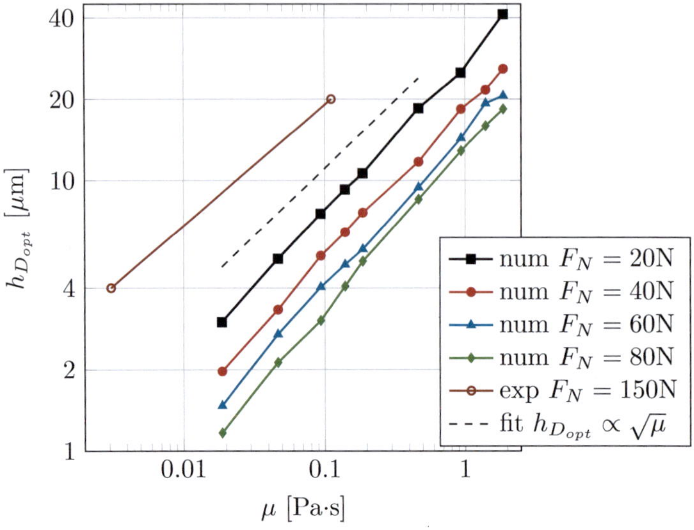

Figure 4.16: Scaling of the optimal dimple depth with respect to the viscosity. Four different values of the normal force are represented. The fitting dashed line corresponds to the scaling shown in equation 4.11. Adapted from [CFM+18].

the depth ones, knowing that the experimental analysis was carried out with constant aspect ratio $\lambda = 10$. As one can see, the slope of the experimental curve is in very good agreement with the simulation results and, hence, with the scaling relationship expressed in equation 4.11.

The relationship expressed in equation 4.11 can be further generalized by introducing the Sommerfeld number as presented in equation 2.8. Moreover, for geometries such as sliding bearings and the one in the current study, Hamrock *et al.* [71] and Raimondi and Boyd [146] introduced a definition of Sommerfeld number, which is based on the so-called shoulder height s_h. The shoulder height is defined as the height difference between the outer edge of the pin and the center of the pin (*i.e.* the point where the gap height has its minimum). The definition of the shoulder height s_h is shown in figure 2.3 for the case of pin *A* and reads $s_h = 6.22$ μm. Such a definition results from the need of defining a fixed geometrical quantity which is comparable to the gap height but, at the same time, it is known *a priori*, since the gap height can freely vary during the simulations. The so-defined Sommerfeld number is hence:

$$\mathcal{S} = \frac{\mu U L_x}{s_h^2 W}. \tag{4.12}$$

By substituting the above definition in equation 4.11 we obtain:

$$h_{D_{opt}} = K\sqrt{\frac{\mu L_x U}{W}} = K\sqrt{\frac{\mu U L s_h^2}{W s_h^2}} = K s_h \sqrt{\mathcal{S}}. \tag{4.13}$$

The shoulder height can be brought to the left hand side in order to define the non-dimensional optimal depth and its dependence on the Sommerfeld number.

$$\frac{h_{D_{opt}}}{s_h} = K\sqrt{\mathcal{S}}. \tag{4.14}$$

This relationship expresses the proportionality of the dimensionless optimal depth to the square root of the Sommerfeld number. Furthermore, the proportionality constant K depends only on the macro geometry of the pin and the arbitrarily chosen reference load W_r.

Figure 4.17 shows the behavior of the dimensionless optimal depth as a function of the Sommerfeld number defined through the shoulder height s_h in equation 4.12. It is interesting to note that all the numerical curves collapse onto a single line, which is, in turn, in good agreement with the proposed scaling (fitting line $h_{D_{opt}} \propto \sqrt{\mathcal{S}}$).

Also in this case, the slope of the experimental results is in very good agreement with the numerical ones, but different in the absolute values. In this case, the reason of this shift can be attributed to two factors. Firstly, the experimental results are obtained in the mixed lubrication regime, meaning that a contribution of the normal force probably comes from the direct contact between the two surfaces or other effects typical of this lubrication regime. Secondly, the

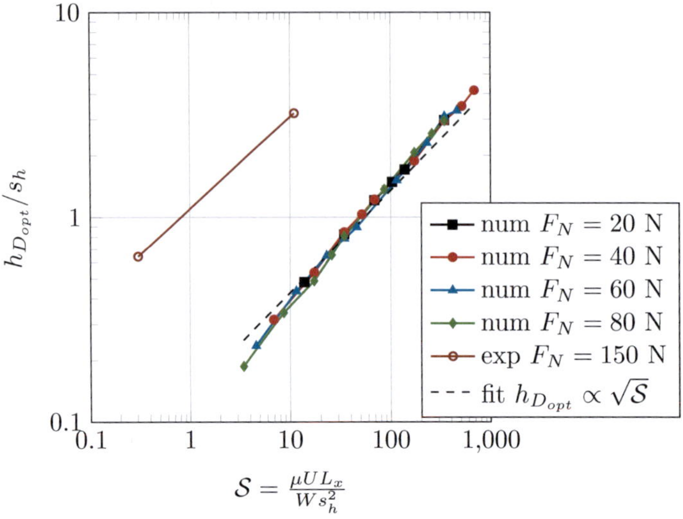

Figure 4.17: Non-dimensional optimal depth as function of the Sommerfeld number. The curves are the same as in the dimensional case shown in figure 4.16. The fitting line represents the scaling proposed in equation 5.2. The experimental results from Braun *et al.* [19] are non-dimensionalized through a constant velocity $U = 1$ m/s. Adapted from [CFM+18].

Sommerfeld number for the experimental values shown in figure 4.17 can be arbitrarily shifted according to the velocity used in the definition since the experimental values of the optimal depth have been measured for different values of velocity. This second aspect will be further investigated in future works.

As conclusion, the overall good agreement in the slope of the experimental and numerical results indicates that the physical mechanisms which are responsible for the scaling of the optimal depth can be successfully investigated with a hydrodynamic approach, even though the experiments are carried out in the mixed lubrication region.

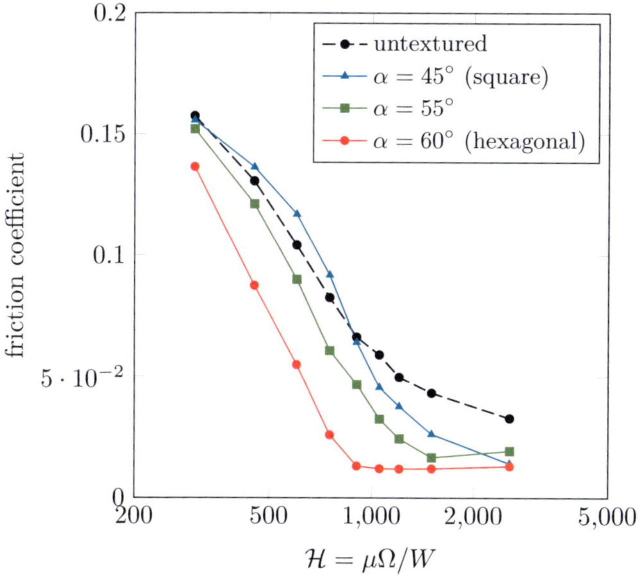

Figure 4.18: Experimental Stribeck curve with different orientations of the texture surface. This experimental results are presented by Schneider *et al.* in [20]. The methodology and the operating conditions of this experimental campaign are espoused in section 2.1. The considered texture has the following parameters: $D = 40\ \mu$m, aspect ratio $\lambda = 10$, texture density $\rho_{txt} = 10\%$. The Stribeck curve is plotted as a function of the Hersey number $\mathcal{H}$. Adapted from [CFM$^+$18] and [20].

4.5 Impact of the dimple distribution

The last parametric investigation concerns the effects of different dispositions of the texture elements. This analysis is motivated by the experimental work of Schneider *et al.* [20], whose results are exposed in figure 4.18. This figure shows the experimental Stribeck curve obtained for different orientations of the texture pattern, ranging from a square one to a pseudo-hexagonal one according to the sketch shown in figure 4.2. As one can see, the pseudo-hexagonal arrangement has an overall lower friction than the other tested arrangements.

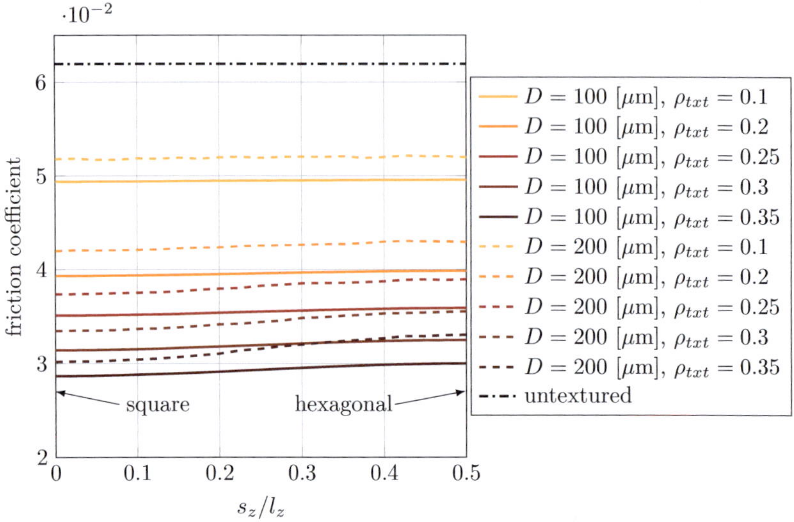

Figure 4.19: Friction coefficient as function of the normalized shift parameter s_z of the texture arrangement. Two groups of curves are presented showing the friction performance for both investigated dimple diameters. For each group five values of texture density are shown. The untextured case is also plotted as reference. The simulation results shown in this figure are obtained under the following operating conditions: $\mu = 0.187$ Pa·s, $U = 0.1$ m/s, $F_N = 15$ N.

The corresponding numerical study is performed by varying the shift parameter s_z (see figure 4.2) in order to generate a set of textures with the same kind of spherical dimple but with a different arrangement as done in the experimental campaign. Moreover, a parameter space with different viscosity and velocity values is considered. The simulations are carried out in the framework of a constant applied load which is set to $F_N = 15$ N as in the experiments. In this case, also the effects of different values of texture density is analyzed, ranging from $\rho_{txt} = 10\%$ to $\rho_{txt} = 35\%$. Two values of the dimple diameter are investigated (*i.e.* $D = 100$ and $D = 200$ μm) while considering a constant aspect ratio $\lambda = 10$.

Figure 4.19 shows the friction coefficient as a function of the shift parameter s_z for two groups of simulations with different dimple diameters. As one can

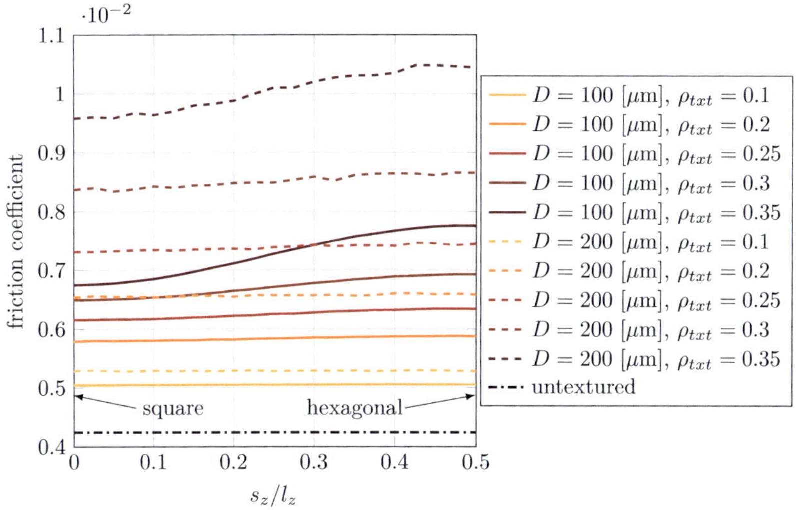

Figure 4.20: Friction coefficient as function of the normalized shift parameter s_z of the texture arrangement. Two groups of curves are presented showing the friction performance for both investigated dimple diameters. For each group five values of texture density are shown. The untextured case is also plotted as reference. The simulation results shown in this figure are obtained under the following operating conditions: $\mu = 0.0187$ Pa·s, $U = 0.1$ m/s, $F_N = 15$ N.

see, for small values of the texture density $\rho_{txt} < 20\%$, the different arrangement of the texture elements does not really impact on the friction coefficient, since the curves remain almost perfectly horizontal regardless of the value of the shift s_z. The influence of the shift parameter becomes more evident for increasing values of the texture density and this influence can clearly be noticed for $\rho_{txt} = 35\%$. Nonetheless, the effects of the texture arrangement appear to act differently than in the experiments shown in figure 4.18. As a matter of fact, the simulations show a lower friction coefficient for the square arrangements than for the pseudo-hexagonal one.

The same trend was found for all the other combinations of velocity and viscosity which were tested in this parametric study. Figure 4.20 shows similar numerical results to those in figure 4.19 but with a lower viscosity $\mu = 0.0187$.

In this case, the textured surface has a worse friction performance than the untextured one. Moreover, also in this case the square pattern shows a lower friction coefficient than the hexagonal one. More generally, the increase of the texture density corresponds to an enhancement of the texture performance in both possible cases, regardless of whether the friction is reduced or increased.

In conclusion, it appears that the influence of the texture disposition cannot be fully represented by means of a solely hydrodynamic approach. The reason behind this discrepancy could lie either in the neglection of the inertial terms (which might describe the possible mutual effects between dimples [80, 128]) or in the absence of contact mechanics modeling. Therefore, despite the implementation of a mass conserving algorithm, the Reynolds equation should be also coupled with a contact mechanic algorithm in order to study the impact of the texture distribution.

5 Conclusion and Outlook

The numerical representation of a pin-on-disc set-up in the hydrodynamic regime is undertaken in this work with the aim to deepen the study of surface texturing as a promising way to reduce friction and wear in lubricated contacts. In order to achieve this goal, this work is divided into two main steps: the first one focuses on the modeling strategies for the description of the hydrodynamic flow in a pin-on-disc tribometer, while the second part concerns the systematic study of the texture design characteristics as a function of the operating conditions. Both parts are exposed in the present work as a homogeneous adaptation based on the three publications of the author [GMB$^+$15, CSMF17, CFM$^+$18] with also extensions and further investigations.

The numerical analysis is based on two main numerical models: the Navier-Stokes equations and the Reynolds equation. These two models are analyzed in both their compressible and incompressible formulation. Moreover, the Reynolds equation is further developed through the addition of a mass-conserving cavitation algorithm.

As first step, a comprehensive comparison between the various numerical models is carried out in order to understand the level of detail that can be achieved in describing the physical phenomena occurring during pin-on-disc investigations in the hydrodynamic lubrication. In this regard, the analysis concerned the numerical description of the pin-on-disc tribometer and its setting parameters, such as the sliding distance, *i.e.* the distance between the pin and the center of rotation of the disc. This analysis is performed with the compressible Navier-Stokes equations and it is based on the real pin geometry as measured from the experimental set-up, so that the simulations can represent the whole system in a way which is as complete as possible. As a result, a dependence is found between the sliding distance and the measured friction coefficient. In particular, the friction coefficient is found to reduce if the pin is moved closer to the center of rotation as also experimentally confirmed by Greiner *et al.* in [GMB$^+$15].

The analysis with the compressible Navier-Stokes equation also brought to light that a faster computational approach is needed in order to run simulations more efficiently considering a richer parameter space. For this reason, the Reynolds equation is considered as a more suitable numerical approach for efficient simulations. In this regard, the applicability hypotheses of the Reynolds equation are examined in the case of a geometry consisting of a pin-on-disc tribometer both at the macro scale of the pin as well as at the micro scale of the texture elements.

For the analysis in the macro scale, several simulations of the pin are carried out with the Navier-Stokes equations for different values of gap height and viscosity with the aim to evaluate the relevance of the inertial terms for a broad range of operating conditions. The contribution of every single term of the Navier-Stokes equation is evaluated in the domain, showing that for the macro geometry in the pin scale, the inertial terms are always negligible in comparison to the pressure and viscous terms. Hence, the analysis through the Reynolds equation could represent a sufficiently good approximation in order to obtain a faster numerical representation of the flow in the macro geometry. The approach used to check the validity of the Reynolds equation in the micro scale which characterize the texture elements follows the scheme proposed by Dobrica and Fillon in [46]. In their work, these authors analyzed the role of the Reynolds number and the role of the aspect ratio of a 2D square dimple by comparing the pressure distribution resulting from the Navier-Stokes equation and the Reynolds one. This work extended that analysis also for 3D spherical texture elements, showing that a slightly lower applicability margin is found with respect to the maximum allowed Reynolds number.

The modeling part of the tribometer set-up is then further analyzed with a particular focus on the relevance of thermodynamic effects in the lubricant film. In particular, the effects of viscous dissipation are investigated in order to provide a characterization of the pressure losses and of the temperature increase as a function of the operating conditions. In order to concentrate the attention rather on the thermodynamic aspect than on the particular feature of the pin-on-disc geometry, the geometry of a parallel channel is considered. The numerical analysis is carried out both with the compressible Navier-Stokes equation and the Reynolds one. The latter is coupled with the energy equation through a generic and versatile formulation which allows to deal with the coupled equations in a rather fast and efficient manner. At first, the applicability region of this approach is identified as a function of the Reynolds and of the Péclet number, respectively $Re < 0.1$ and $Pe < 10^3$. The thereby validated model is then applied to the study of the pressure losses as function of four operating parameters, such as viscosity μ, velocity of the moving wall U, reference length of the bearing L_x and gap height H. These four parameters can be resumed through the introduction of the Sommerfeld number defined as $\mathcal{S} = UL_x\mu/(H^2 p_r)$, where p_r is the reference pressure. It was found that, when the Sommerfeld number increases, the temperature distribution increases as well along the channel due to the viscous dissipation. This temperature increase affects also the density and the pressure distribution. In particular, it is shown that the pressure distribution has to decrease (pressure loss) along the channel in order to counterbalance the increase of thermal energy in the flow through a decrease of mechanical energy. This mechanism was highlighted through the analysis of the contribution of every term in the energy equation. As quantitative results, the mechanical energy losses become relevant for $\mathcal{S} > 1$. Moreover, it was shown that the pressure losses scale with the square of the Sommerfeld number while the inlet-outlet temperature difference increases linearly with the Sommerfeld number.

In the second part of this work, the geometry of the tribometer is considered also in the presence of a texture on the pin surface. The numerical approach is based on the incompressible Reynolds equation with mass-conserving cavitation, whereas particular emphasis is laid on the computational performance of the solver in order to carry out parametric studies based on high-resolution surface representations of experimental set-ups. Three main studies are undertaken with the goal of investigating the texture performance focusing on the relations and scalings between operating parameters and the optimal shape of dimples. The first study concerns the sensitivity of the load carrying capacity with respect to the dimple position on the pin surface. Consistently to the literature results and the experimental results shown in [CFM+18], it is found that the front part of the pin surface experiences the best benefits from the presence of a texture, while on the rear part of the pin, where cavitation occurs, the texture elements have either no influence or even a detrimental effect. Moreover, this analysis showed that for simulations with prescribed gap height, viscosity and velocity have a much smaller influence than the gap height in defining the position in which dimples have a positive effect on the normal force.

The second study of the textured surface focuses on the influence of the operating parameters on the dimple diameter and depth. At first, the load carrying capacity is found to have a maximum in the depth-diameter space which is independent of the diameter value. This hints to the fact that an optimal value of the dimple depth exists which is shown to be a linear function of the gap height. Starting from the generic function which describes the optimal depth $h_{D_{opt}} = h_{D_{opt}}(\mu, U, H, D, W)$, the linear relationship between the optimal depth and the gap height can be expressed in the following way:

$$h_{D_{opt}} = kH(\mu, U, W) = K\sqrt{\frac{\mu L_x U}{W}} \tag{5.1}$$

where μ is the viscosity, L_x the reference length, U the sliding velocity, W the applied load and D the diameter. The latter dependence is excluded from equation 5.1 since the distribution of the maximal normal force in the depth-diameter space is independent of the dimple diameter.

A further generalization of the scaling results comes from the introduction of the shoulder height s_h as reference length for the gap height [71, 146]. This allows to rewrite the above equation 5.1 by considering the Sommerfeld number as defined in equation 4.12 and, hence, leading to:

$$\frac{h_{D_{opt}}}{s_h} = K\sqrt{S}. \tag{5.2}$$

This equation shows that the dimensionless optimal dimple depth is a function of a constant K (which depends only on the shape of the macro geometry and on the arbitrary reference load W_r) times the square root of the Sommerfeld number. This scaling has been proven through simulations which employed both possible simulating strategies, namely with prescribed gap height and with constant normal force. The scaling results are also in very good agreement with the experimental results presented by Braun *et al.* in [19], suggesting that hydrodynamic mechanics induced by dimples also play a relevant role in the mixed lubrication region.

The third parametric analysis concerns the role of the texture orientation on the surface. The numerical analysis in the hydrodynamic regime showed that for low values of the texture density almost no influence is noticed between different kind of dimple dispositions. On the other side, for a relatively high texture density the square arrangement of dimples appears to have a lower friction coefficient than in the case with a pseudo-hexagonal pattern. These findings are in contrast with those of the experimental analysis performed by Schneider *et al.* [20] where the pseudo-hexagonal pattern is found to perform better than the square one. The reason for this discrepancy may lie in the fact that the simulations are carried out in the hydrodynamic regime while the experiments were performed under mixed lubrication conditions. Hence, the fluid dynamic model based on the Reynolds equation with mass-conserving cavitation has to be complemented with the addition of models which can consider the influence of the other physical phenomena which occur in the mixed lubrication regime.

The consideration of the coupling between the Reynolds equation and the complementary models represents the main outlook for the present work. In this regard, the aim would be the progressive shift of focus from the hydrodynamic region to the mixed regime. This could be done by also considering models for the interaction between the roughness and the fluid dynamics (such as flow factors [147, 148] or homogenization techniques [149, 150]) and also the interaction between the pressure in the fluid film and the mechanical deformation of the surfaces through a fully elasto-hydrodynamic approach [63, 151, 152]. In addition, the above-mentioned steps can be coupled with a contact mechanics algorithm which can describe the resulting interaction between the two surfaces as soon as they get in contact [153, 154]. All the above-mentioned methods can be finally coupled in a fully thermo-elasto-hydrodynamic approach [145, 155] paying attention to an efficient implementation so that a systematic optimization of the textured surfaces can be carried out.

Bibliography

[1] D. Dowson. *History of tribology*. Addison-Wesley Longman Limited, London, 1979.

[2] B. Bhushan. *Introduction to tribology*. John Wiley & Sons, New York, 2013.

[3] H. P. Jost. *Lubrication: Tribology; Education and Research; Report on the Present Position and Industry's Needs (submitted to the Department of Education and Science by the Lubrication Engineering and Research) Working Group*. HM Stationery Office, 1966.

[4] H. P. Jost. Tribology, origin and future. *Wear*, 136(1):1–17, 1990.

[5] C. Evans. Strategy for energy conservation through tribology. *American Society of Mechanical Engineers, New York*, page 174, 1978.

[6] P. Jost. Economic impact of tribology. *Proc Mechanical Failures Prevention Group*, pages 117–139, 1976.

[7] R. W. Carpick, A. Jackson, W. G. Sawyer, N. Argibay, P. Lee, A. Pachon, and R. M. Gresham. The tribology opportunities study: Can tribology save a quad? *Tribology & Lubrication Technology*, 72(5):44, 2016.

[8] O. Pinkus and D. F. Wilcock. Strategy for energy conservation through tribology. *American Society of Mechanical Engineers, New York*, 1977.

[9] K. Holmberg, P. Kivikytö-Reponen, P. Härkisaari, K. Valtonen, and A. Erdemir. Global energy consumption due to friction and wear in the mining industry. *Tribology International*, 115:116–139, 2017.

[10] A. E Somers, P. C. Howlett, D. R. MacFarlane, and M. Forsyth. A review of ionic liquid lubricants. *Lubricants*, 1(1):3–21, 2013.

[11] T.W. Scharf and S.V. Prasad. Solid lubricants: a review. *Journal of materials science*, 48(2):511–531, 2013.

[12] N. Shaigan, W. Qu, D. G. Ivey, and W. Chen. A review of recent progress in coatings, surface modifications and alloy developments for solid oxide fuel cell ferritic stainless steel interconnects. *Journal of Power Sources*, 195(6):1529–1542, 2010.

[13] S. Kalpakjian and S. R. Schmid. *Manufacturing engineering and technology*. Pearson Upper Saddle River, NJ, USA, 2014.

[14] S. C. Tung and M. L. McMillan. Automotive tribology overview of current advances and challenges for the future. *Tribology International*, 37(7):517–536, 2004.

[15] K. Bobzin, T. Bartels, et al. *Industrial tribology: Tribosystems, friction, wear and surface engineering, lubrication*. John Wiley & Sons, Hoboken (NJ), 2011.

[16] D. Gropper, L. Wang, and T. J. Harvey. Hydrodynamic lubrication of textured surfaces: A review of modeling techniques and key findings. *Tribology International*, 94:509–529, 2016.

[17] I. Etsion. State of the art in laser surface texturing. *Journal of tribology*, 127(1):248–253, 2005.

[18] I. Etsion. Modeling of surface texturing in hydrodynamic lubrication. *Friction*, 1(3):195–209, 2013.

[19] D. Braun, C. Greiner, J. Schneider, and P. Gumbsch. Efficiency of laser surface texturing in the reduction of friction under mixed lubrication. *Tribology International*, 77:142–147, 2014.

[20] D. Braun J. Schneider and C. Greiner. Laser textured surfaces for mixed lubrication: Influence of aspect ratio, textured area and dimple arrangement. *Lubricants*, 5(3):32, 2017.

[21] V. Brizmer, Y. Kligerman, and I. Etsion. A laser surface textured parallel thrust bearing. *Tribology Transactions*, 46(3):397–403, 2003.

[22] I. Etsion, G. Halperin, V. Brizmer, and Y. Kligerman. Experimental investigation of laser surface textured parallel thrust bearings. *Tribology Letters*, 17(2):295–300, 2004.

[23] X. Wang, K. Kato, K. Adachi, and K. Aizawa. Loads carrying capacity map for the surface texture design of sic thrust bearing sliding in water. *Tribology International*, 36(3):189–197, 2003.

[24] Y. Henry, J. Bouyer, and M. Fillon. Experimental analysis of the hydrodynamic effect during start-up of fixed geometry thrust bearings. *Tribology International*, 229(4):362–375, 2017.

[25] D. Yan, N. Qu, H. Li, and X. Wang. Significance of dimple parameters on the friction of sliding surfaces investigated by orthogonal experiments. *Tribology Transactions*, 53(5):703–712, 2010.

[26] X. Lu and M.M. Khonsari. An experimental investigation of dimple effect on the stribeck curve of journal bearings. *Tribology Letters*, 27(2):169–176, 2007.

[27] M.J. Conlon, A. Dadouche, W.M. Dmochowski, R. Payette, J. Bédard, and B. Liko. Experimental evaluation of foil bearing performance: Steady-state and dynamic results. In *ASME Turbo Expo 2009: Power*

for Land, Sea, and Air, pages 981–992. American Society of Mechanical Engineers, 2009.

[28] L. Bai and S. Bai. Frictional performance of a textured surface with elliptical dimples: geometric and distribution effects. *Tribology Transactions*, 57(6):1122–1128, 2014.

[29] X.Q. Yu, S. He, and R.L. Cai. Frictional characteristics of mechanical seals with a laser-textured seal face. *Journal of Materials Processing Technology*, 129(1-3):463–466, 2002.

[30] G. Ryk, Y. Kligerman, and I. Etsion. Experimental investigation of laser surface texturing for reciprocating automotive components. *Tribology Transactions*, 45(4):444–449, 2002.

[31] F.J. Profito, S. Vlădescu, T. Reddyhoff, and D. Dini. Experimental validation of a mixed-lubrication regime model for textured piston-ring-liner contacts. *Materials Performance and Characterization*, 6(2):112–129, 2017.

[32] S. Vlădescu, A. V. Olver, I. G. Pegg, and T. Reddyhoff. The effects of surface texture in reciprocating contacts–an experimental study. *Tribology International*, 82:28–42, 2015.

[33] M. Wakuda, Y. Yamauchi, S. Kanzaki, and Y. Yasuda. Effect of surface texturing on friction reduction between ceramic and steel materials under lubricated sliding contact. *Wear*, 254(3-4):356–363, 2003.

[34] M. Nakano, A. Korenaga, A. Korenaga, K. Miyake, T. Murakami, Y. Ando, H. Usami, and S. Sasaki. Applying micro-texture to cast iron surfaces to reduce the friction coefficient under lubricated conditions. *Tribology Letters*, 28(2):131–137, 2007.

[35] A. Borghi, E. Gualtieri, D. Marchetto, L. Moretti, and S. Valeri. Tribological effects of surface texturing on nitriding steel for high-performance engine applications. *Wear*, 265(7-8):1046–1051, 2008.

[36] A. Kovalchenko, O. Ajayi, A. Erdemir, G. Fenske, and I. Etsion. The effect of laser surface texturing on transitions in lubrication regimes during unidirectional sliding contact. *Tribology International*, 38(3):219–225, 2005.

[37] H. Zhang, D.Y. Zhang, M. Hua, G.N. Dong, and K.S. Chin. A study on the tribological behavior of surface texturing on babbitt alloy under mixed or starved lubrication. *Tribology Letters*, 56(2):305–315, 2014.

[38] M. Scaraggi, F. P. Mezzapesa, G. Carbone, A. Ancona, and L. Tricarico. Friction properties of lubricated laser-microtextured-surfaces: an experimental study from boundary-to hydrodynamic-lubrication. *Tribology Letters*, 49(1):117–125, 2013.

[39] M. Scaraggi, F. P. Mezzapesa, G. Carbone, A. Ancona, D. Sorgente, and P. M. Lugarà. Minimize friction of lubricated laser-microtextured-surfaces by tuning microholes depth. *Tribology International*, 75:123–127, 2014.

[40] D.B. Hamilton, J.A. Walowit, and C.M. Allen. A theory of lubrication by microirregularities. *Journal of Basic Engineering*, 88(1):177–185, 1966.

[41] H. Yamakiri, S. Sasaki, T. Kurita, and N. Kasashima. Effects of laser surface texturing on friction behavior of silicon nitride under lubrication with water. *Tribology International*, 44(5):579–584, 2011.

[42] M. Varenberg, G. Halperin, and I. Etsion. Different aspects of the role of wear debris in fretting wear. *Wear*, 252(11):902–910, 2002.

[43] Y. Wan and D. Xiong. The effect of laser surface texturing on frictional performance of face seal. *Journal of Materials Processing Technology*, 197(1):96–100, 2008.

[44] C. Greiner, M. Schäfer, U. Popp, and P. Gumbsch. Contact splitting and the effect of dimple depth on static friction of textured surfaces. *ACS applied materials & interfaces*, 6(11):7986–7990, 2014.

[45] M. Arghir, N. Roucou, M. Helene, and J. Frene. Theoretical analysis of the incompressible laminar flow in a macro-roughness cell. *Journal of tribology*, 125(2):309–318, 2003.

[46] M.B. Dobrica and M. Fillon. About the validity of Reynolds equation and inertia effects in textured sliders of infinite width. *Proceedings of the Institution of Mechanical Engineers, Part J: Journal of Engineering Tribology*, 223(1):69–78, 2009.

[47] M.J. Braun and W.M. Hannon. Cavitation formation and modelling for fluid film bearings: a review. *Proceedings of the Institution of Mechanical Engineers, Part J: Journal of Engineering Tribology*, 224(9):839–863, 2010.

[48] C.I. Papadopoulos, L. Kaiktsis, and M. Fillon. Computational fluid dynamics thermohydrodynamic analysis of three-dimensional sector-pad thrust bearings with rectangular dimples. *Journal of Tribology*, 136(1):011702, 2014.

[49] M.D. Pascovici, T. Cicone, M. Fillon, and M.B. Dobrica. Analytical investigation of a partially textured parallel slider. *Proceedings of the Institution of Mechanical Engineers, Part J: Journal of Engineering Tribology*, 223(2):151–158, 2009.

[50] A. Ramesh, W. Akram, S. P. Mishra, A. H. Cannon, A. A. Polycarpou, and W. P. King. Friction characteristics of microtextured surfaces under

mixed and hydrodynamic lubrication. *Tribology International*, 57:170–176, 2013.

[51] H. Yu, W. Huang, and X. Wang. Dimple patterns design for different circumstances. *Lubrication Science*, 25(2):67–78, 2013.

[52] N. Morris, M. Leighton, M. De la Cruz, R. Rahmani, H. Rahnejat, and S. Howell-Smith. Combined numerical and experimental investigation of the micro-hydrodynamics of chevron-based textured patterns influencing conjunctional friction of sliding contacts. *Proceedings of the Institution of Mechanical Engineers, Part J: Journal of Engineering Tribology*, 229(4):316–335, 2015.

[53] R. Wahl, J. Schneider, and P. Gumbsch. In situ observation of cavitation in crossed microchannels. *Tribology International*, 55:81–86, 2012.

[54] B. A. Krick, J. R Vail, B. N. J. Persson, and W. G. Sawyer. Optical in situ micro tribometer for analysis of real contact area for contact mechanics, adhesion, and sliding experiments. *Tribology Letters*, 45(1):185–194, 2012.

[55] E. M. Bortoleto, A. C. Rovani, V. Seriacopi, F. J. Profito, D. C. Zachariadis, I. F. Machado, A. Sinatora, and R. M. Souza. Experimental and numerical analysis of dry contact in the pin on disc test. *Wear*, 301(1-2):19–26, 2013.

[56] W. Yan, N. P. O'Dowd, and E. P. Busso. Numerical study of sliding wear caused by a loaded pin on a rotating disc. *Journal of the Mechanics and Physics of Solids*, 50(3):449–470, 2002.

[57] N. Laraqi, N. Alilat, J.M. Garcia de Maria, and A. Bairi. Temperature and division of heat in a pin-on-disc frictional device-exact analytical solution. *Wear*, 266(7-8):765–770, 2009.

[58] D. Dowson and G. R. Higginson. *Elasto-hydrodynamic lubrication: international series on materials science and technology*, volume 23. Elsevier, Pergamon, 2014.

[59] D. Dowson, G.R. Higginson, and A.V. Whitaker. Elasto-hydrodynamic lubrication: a survey of isothermal solutions. *Journal of Mechanical Engineering Science*, 4(2):121–126, 1962.

[60] A.A. Lubrecht, W.E. Ten Napel, and R. Bosma. Multigrid, an alternative method for calculating film thickness and pressure profiles in elastohydrodynamically lubricated line contacts. *Journal of tribology*, 108(4):551–556, 1986.

[61] C. H. Venner and A. A. Lubrecht. *Multi-level methods in lubrication*, volume 37. Elsevier, Amsterdam, 2000.

[62] T. Doki-Thonon, N. Fillot, G. Espejel, M. Querry, D. Philippon, N. Devaux, and P. Vergne. A dual experimental/numerical approach for film thickness analysis in tehl spinning skewing circular contacts. *Tribology Letters*, 50(1):115–126, 2013.

[63] W. Habchi, D. Eyheramendy, P. Vergne, and G. Morales-Espejel. A full-system approach of the elastohydrodynamic line/point contact problem. *Journal of Tribology*, 130(2):021501, 2008.

[64] W. Habchi, D. Eyheramendy, P. Vergne, and G. Morales-Espejel. Stabilized fully-coupled finite elements for elastohydrodynamic lubrication problems. *Advances in Engineering Software*, 46(1):4–18, 2012.

[65] H.J. Kim, P. Ehret, D. Dowson, and C.M. Taylor. Thermal elastohydrodynamic analysis of circular contacts part 1: Newtonian model. *Proceedings of the Institution of Mechanical Engineers, Part J: Journal of Engineering Tribology*, 215(4):339–352, 2001.

[66] H.J. Kim, P. Ehret, D. Dowson, and C.M. Taylor. Thermal elastohydrodynamic analysis of circular contacts part 2: Non-newtonian model. *Proceedings of the Institution of Mechanical Engineers, Part J: Journal of Engineering Tribology*, 215(4):353–362, 2001.

[67] V. Bruyere, N. Fillot, G. E Morales-Espejel, and P. Vergne. Computational fluid dynamics and full elasticity model for sliding line thermal elastohydrodynamic contacts. *Tribology International*, 46(1):3–13, 2012.

[68] H. G. Elrod. A cavitation algorithm. *ASME J. Lubr. Technol.*, 103:350, 1981.

[69] T. Woloszynski, P. Podsiadlo, and G. W. Stachowiak. Efficient solution to the cavitation problem in hydrodynamic lubrication. *Tribology Letters*, 58(1):18, 2015.

[70] R. Stribeck. Characteristics of plain and roller bearings. *Zeit. VDI*, 46, 1902.

[71] B. J. Hamrock, S. R. Schmid, and B. O. Jacobson. *Fundamentals of fluid film lubrication*. Marcel Dekker, Inc., New York, 2004.

[72] M. Scaraggi. Textured surface hydrodynamic lubrication: discussion. *Tribology Letters*, 48(3):375–391, 2012.

[73] K. Zum Gahr, R. Wahl, and K. Wauthier. Experimental study of the effect of microtexturing on oil lubricated ceramic/steel friction pairs. *Wear*, 267(5):1241–1251, 2009.

[74] J Dumond, F Magagnato, and A Class. Stochastic-field cavitation model. *Physics of Fluids*, 25(7):073302, 2013.

[75] E. Turkel. Review of preconditioning methods for fluid dynamics. *Applied Numerical Mathematics*, 12(1-3):257–284, 1993.

[76] J. H. Ferziger and M. Peric. *Computational methods for fluid dynamics*. Springer Science & Business Media, Berlin, 2012.

[77] ANSYS corporation. Ansys academic research mechanical, release 16.1, 2015.

[78] D. Dowson. A generalized Reynolds equation for fluid-film lubrication. *International Journal of Mechanical Sciences*, 4(2):159–170, 1962.

[79] E. Guyon, J. Hulin, L. Petit, and C. D. Mitescu. *Physical hydrodynamics*. Oxford University Press, Oxford, 2015.

[80] T. Woloszynski, P. Podsiadlo, and G. W. Stachowiak. Evaluation of inertia effect in finite hydrodynamic bearings with surface texturing using spectral element solver. *Tribology International*, 91:170–176, 2015.

[81] M. Arghir, A. Alsayed, and D. Nicolas. The finite volume solution of the Reynolds equation of lubrication with film discontinuities. *International journal of mechanical sciences*, 44(10):2119–2132, 2002.

[82] M.M. Khonsari and J.J. Beaman. Thermohydrodynamic analysis of laminar incompressible journal bearings. *ASLE transactions*, 29(2):141–150, 1986.

[83] Josiah D. Knight and L.E. Barrett. Analysis of axially grooved journal bearings with heat transfer effects. *ASLE transactions*, 30(3):316–323, 1987.

[84] W. Habchi and S. Bair. Quantitative compressibility effects in thermal elastohydrodynamic circular contacts. *Journal of Tribology*, 135(1):011502, 2013.

[85] T. Doki-Thonon, N. Fillot, P. Vergne, and G.E. M. Espejel. Numerical insight into heat transfer and power losses in spinning ehd

non-newtonian point contacts. *Proceedings of the institution of mechanical engineers, Part J: Journal of Engineering Tribology*, page 1350650111419567, 2011.

[86] S. Kango, R.K. Sharma, and R.K. Pandey. Thermal analysis of microtextured journal bearing using non-newtonian rheology of lubricant and jfo boundary conditions. *Tribology International*, 69:19–29, 2014.

[87] H.G. Elrod. A computer program for cavitation and starvation problems. *Cavitation and related phenomena in lubrication*, 37, 1974.

[88] B. Jakobsson. The finite journal bearing considering vaporization. *Trans. Chalmers Univ. of Tech, Sweden*, 190, 1965.

[89] K. Olsson. Cavitation in dynamically loaded bearing. *Trans. Chalmers Univ. of Tech, Sweden*, 308, 1957.

[90] F. Sahlin, A. Almqvist, R. Larsson, and S. Glavatskih. A cavitation algorithm for arbitrary lubricant compressibility. *Tribology International*, 40(8):1294–1300, 2007.

[91] L. Bertocchi, D. Dini, M. Giacopini, M. T Fowell, and A. Baldini. Fluid film lubrication in the presence of cavitation: a mass-conserving two-dimensional formulation for compressible, piezoviscous and non-newtonian fluids. *Tribology International*, 67:61–71, 2013.

[92] C. T. Kelley. *Solving nonlinear equations with Newton's method*, volume 1. Siam, Philadelphia, 2003.

[93] Y. Kunibe and H. Tamai. Automatic engine stop-start system, August 5 1997. US Patent 5,653,659.

[94] G. Stachowiak and A. W. Batchelor. *Engineering tribology*. Butterworth-Heinemann, Oxford, 2013.

[95] R. D. Moser, J. Kim, and N. N. Mansour. Direct numerical simulation of turbulent channel flow up to $re_{\tau} = 590$. *Physics of fluids*, 11(4):943–945, 1999.

[96] R. Ćerimagić, P. Johansen, T. O. Andersen, and H. C. Pedersen. State of the art review on theoretical tribology of fluid power displacement machines. In *9th FPNI Ph. D. Symposium on Fluid Power*, pages V001T01A026–V001T01A026. American Society of Mechanical Engineers, 2016.

[97] A. Cameron. The viscosity wedge. *ASLE transactions*, 1(2):248–253, 1958.

[98] W.F. Cope. The hydrodynamical theory of film lubrication. *Proc. R. Soc. Lond. A*, 197(1049):201–217, 1949.

[99] R.K. Sharma and R.K. Pandey. Thermohydrodynamic analysis of infinitely wide cycloidal profiled pad thrust bearing with rough surface and a comparison to plane profiled pad. *Lubrication Science*, 20(3):183–203, 2008.

[100] D. Bonneau, A. Fatu, and D. Souchet. *Thermo-hydrodynamic lubrication in hydrodynamic bearings*. John Wiley & Sons, Hoboken (NJ), 2014.

[101] P.M. Lugt and G. E. Morales-Espejel. A review of elasto-hydrodynamic lubrication theory. *Tribology Transactions*, 54(3):470–496, 2011.

[102] Y. Okamoto. Numerical analysis of lubrication in a journal bearing by a thermoelastohydrodynamic lubrication (tehl) model. *International Journal of Engine Research*, 6(2):95–105, 2005.

[103] S. Kim and I. Mudawar. Review of databases and predictive methods for pressure drop in adiabatic, condensing and boiling mini/micro-channel flows. *International Journal of Heat and Mass Transfer*, 77:74–97, 2014.

[104] W. Lee, M. Wong, and Y. Zohar. Pressure loss in constriction microchannels. *Microelectromechanical Systems, Journal of*, 11(3):236–244, 2002.

[105] S.J. Chowdhury and G. Ahmadi. Thermodynamic analysis of wide thrust bearings operating in laminar inertial flow regimes. *Tribology International*, 19(6):281–288, 1986.

[106] W. Habchi, P. Vergne, S. Bair, O. Andersson, D. Eyheramendy, and G. E. Morales-Espejel. Influence of pressure and temperature dependence of thermal properties of a lubricant on the behaviour of circular tehd contacts. *Tribology International*, 43(10):1842–1850, 2010.

[107] M.M. Khonsari. A review of thermal effects in hydrodynamic bearings. part ii: journal bearings. *ASLE transactions*, 30(1):26–33, 1987.

[108] H. S. Hsiao and B. J. Hamrock. A complete solution for thermal-elastohydrodynamic lubrication of line contacts using circular non-newtonian fluid model. *Journal of tribology*, 114(3):540–551, 1992.

[109] J.P. Newall, S. Cowperthwaite, M. Hough, and A.P. Lee. Efficiency modelling in the full toroidal variator: Investigation into optimisation of ehl contact conditions to maximize contact efficiency. *Tribology and Interface Engineering Series*, 48:245–255, 2005.

[110] T. Regueira, L. Lugo, O. Fandiño, E. R. López, and J. Fernández. Compressibilities and viscosities of reference and vegetable oils for their use as hydraulic fluids and lubricants. *Green Chemistry*, 13(5):1293–1302, 2011.

[111] R. Saurel and R. Abgrall. A simple method for compressible multifluid flows. *SIAM Journal on Scientific Computing*, 21(3):1115–1145, 1999.

[112] P.G. Huang, G.N. Coleman, and P. Bradshaw. Compressible turbulent channel flows: Dns results and modelling. *Journal of Fluid Mechanics*, 305:185–218, 1995.

[113] P. Andersson, J. Koskinen, S. Varjus, Y. Gerbig, H. Haefke, S. Georgiou, B. Zhmud, and W. Buss. Microlubrication effect by laser-textured steel surfaces. *Wear*, 262(3-4):369–379, 2007.

[114] U. Pettersson and S. Jacobson. Influence of surface texture on boundary lubricated sliding contacts. *Tribology International*, 36(11):857–864, 2003.

[115] S. Kobatake, Y. Kawakubo, and S. Suzuki. Laplace pressure measurement on laser textured thin-film disk. *Tribology international*, 36(4-6):329–333, 2003.

[116] A. Shinkarenko, Y. Kligerman, and I. Etsion. Theoretical analysis of surface-textured elastomer sleeve in lubricated rotary sliding. *Tribology Transactions*, 53(3):376–385, 2010.

[117] Y. Kligerman, I. Etsion, and A. Shinkarenko. Improving tribological performance of piston rings by partial surface texturing. *Journal of Tribology*, 127(3):632–638, 2005.

[118] A. Ronen, I. Etsion, and Y. Kligerman. Friction-reducing surface-texturing in reciprocating automotive components. *Tribology Transactions*, 44(3):359–366, 2001.

[119] A. Shinkarenko, Y. Kligerman, and I. Etsion. The validity of linear elasticity in analyzing surface texturing effect for elastohydrodynamic lubrication. *Journal of Tribology*, 131(2):021503, 2009.

[120] M.T. Fowell, S. Medina, A.V. Olver, H.A. Spikes, and I.G. Pegg. Parametric study of texturing in convergent bearings. *Tribology International*, 52:7–16, 2012.

[121] R. Rahmani, I. Mirzaee, A. Shirvani, and H. Shirvani. An analytical approach for analysis and optimisation of slider bearings with infinite width parallel textures. *Tribology International*, 43(8):1551–1565, 2010.

[122] A. Guzek, P. Podsiadlo, and G. W. Stachowiak. Optimization of textured surface in 2d parallel bearings governed by the Reynolds equation including cavitation and temperature. *Tribology Online*, 8(1):7–21, 2013.

[123] C. Shen and M.M. Khonsari. Numerical optimization of texture shape for parallel surfaces under unidirectional and bidirectional sliding. *Tribology International*, 82:1–11, 2015.

[124] R. Rahmani, A. Shirvani, and H. Shirvani. Optimization of partially textured parallel thrust bearings with square-shaped micro-dimples. *Tribology Transactions*, 50(3):401–406, 2007.

[125] D. Gropper, T. J. Harvey, and L. Wang. A numerical model for design and optimization of surface textures for tilting pad thrust bearings. *Tribology International*, 119:190–207, 2018.

[126] A. Guzek, P. Podsiadlo, and G. W. Stachowiak. A unified computational approach to the optimization of surface textures: one dimensional hydrodynamic bearings. *Tribology Online*, 5(3):150–160, 2010.

[127] X. Wang, J. Wang, B. Zhang, and W. Huang. Design principles for the area density of dimple patterns. *Proceedings of the Institution of Mechanical Engineers, Part J: Journal of Engineering Tribology*, 229(4):538–546, 2015.

[128] M.B. Dobrica, M. Fillon, M.D. Pascovici, and T. Cicone. Optimizing surface texture for hydrodynamic lubricated contacts using a mass-conserving numerical approach. *Proceedings of the Institution of Mechanical Engineers, Part J: Journal of Engineering Tribology*, 224(8):737–750, 2010.

[129] N. Tala-Ighil, M. Fillon, and P. Maspeyrot. Effect of textured area on the performances of a hydrodynamic journal bearing. *Tribology International*, 44(3):211–219, 2011.

[130] N. Tala-Ighil, P. Maspeyrot, M. Fillon, and A. Bounif. Effects of surface texture on journal-bearing characteristics under steady-state operating conditions. *Proceedings of the Institution of Mechanical Engineers, Part J: Journal of Engineering Tribology*, 221(6):623–633, 2007.

[131] S. Vlădescu, A. Ciniero, K. Tufail, A. Gangopadhyay, and T. Reddyhoff. Optimisation of pocket geometry for friction reduction in piston-liner contacts. *Tribology Transactions*, pages 1–10, 2017.

[132] V. Brizmer, Y. Kligerman, and I. Etsion. A laser surface textured parallel thrust bearing. *Tribology Transactions*, 46(3):397–403, 2003.

[133] S. Cupillard, M. J. Cervantes, and S. Glavatskih. Pressure buildup mechanism in a textured inlet of a hydrodynamic contact. *Journal of Tribology*, 130(2):021701, 2008.

[134] H. Yu, H. Deng, W. Huang, and X. Wang. The effect of dimple shapes on friction of parallel surfaces. *Proceedings of the Institution of Mechanical Engineers, Part J: Journal of Engineering Tribology*, 225(8):693–703, 2011.

[135] A. Gherca, A. Fatu, M. Hajjam, and P. Maspeyrot. Effects of surface texturing in steady-state and transient flow conditions: Two-dimensional numerical simulation using a mass-conserving cavitation

model. *Proceedings of the Institution of Mechanical Engineers, Part J: Journal of Engineering Tribology*, 229(4):505–522, 2015.

[136] A.N. Murthy, I. Etsion, and F.E. Talke. Analysis of surface textured air bearing sliders with rarefaction effects. *Tribology Letters*, 28(3):251–261, 2007.

[137] M. Adjemout, N. Brunetiere, and J. Bouyer. Numerical analysis of the texture effect on the hydrodynamic performance of a mechanical seal. *Surface Topography: Metrology and Properties*, 4(1):014002, 2015.

[138] V.G. Marian, M. Kilian, and W. Scholz. Theoretical and experimental analysis of a partially textured thrust bearing with square dimples. *Proceedings of the Institution of Mechanical Engineers, Part J: Journal of Engineering Tribology*, 221(7):771–778, 2007.

[139] Y. Qiu and M.M. Khonsari. Performance analysis of full-film textured surfaces with consideration of roughness effects. *Journal of tribology*, 133(2):021704, 2011.

[140] N. Brunetière and B. Tournerie. Numerical analysis of a surface-textured mechanical seal operating in mixed lubrication regime. *Tribology International*, 49:80–89, 2012.

[141] S. Kango, R.K. Sharma, and R.K. Pandey. Comparative analysis of textured and grooved hydrodynamic journal bearing. *Proceedings of the Institution of Mechanical Engineers, Part J: Journal of Engineering Tribology*, 228(1):82–95, 2014.

[142] H. Zhang, G. Dong, M. Hua, F. Guo, and K. S. Chin. Parametric design of surface textures on journal bearing. *Industrial Lubrication and Tribology*, 67(4):359–369, 2015.

[143] F. Sahlin, S. B. Glavatskih, T. Almqvist, and R. Larsson. Two-dimensional CFD-analysis of micro-patterned surfaces in hydrodynamic lubrication. *Transactions of the ASME-F-Journal of Tribology*, 127(1):96–102, 2005.

[144] Hu Yong and Raj Balendra. Cfd analysis on the lubrication behaviours of journal bearing with dimples. In *Mechatronics and Automation, 2009. ICMA 2009. International Conference on*, pages 1279–1284. IEEE, 2009.

[145] C. Gu, X. Meng, Y. Xie, and D. Zhang. Mixed lubrication problems in the presence of textures: An efficient solution to the cavitation problem with consideration of roughness effects. *Tribology International*, 103:516–528, 2016.

[146] A.A. Raimondi and J. Boyd. Applying bearing theory to the analysis and design of pad-type bearings. *Trans. ASME*, 77(3):287–309, 1955.

[147] N. Patir and H.S. Cheng. An average flow model for determining effects of three-dimensional roughness on partial hydrodynamic lubrication. *Journal of Tribology*, 100(1):12–17, 1978.

[148] N. Patir and H.S. Cheng. Application of average flow model to lubrication between rough sliding surfaces. *Journal of Tribology*, 101(2):220–229, 1979.

[149] H. G. Elrod. Thin-film lubrication theory for newtonian fluids with surfaces possessing striated roughness or grooving. *Journal of Lubrication Technology*, 95(4):484–489, 1973.

[150] R. Larsson. Modelling the effect of surface roughness on lubrication in all regimes. *Tribology International*, 42(4):512–516, 2009.

[151] R. Lee, C. Hsu, and W. Kuo. Multilevel solution for thermal elastohydrodynamic lubrication of rolling/sliding circular contacts. *Tribology International*, 28(8):541–552, 1995.

[152] M. Nabhani, M. El Khlifi, O. S. T. Gbehe, and B. Bou-Saïd. Coupled couple stress and surface roughness effects on elasto-hydrodynamic contact. *Lubrication science*, 26(4):251–271, 2014.

[153] P. Sainsot and A.A. Lubrecht. Efficient solution of the dry contact of rough surfaces: a comparison of fast fourier transform and multigrid methods. *Proceedings of the Institution of Mechanical Engineers, Part J: Journal of Engineering Tribology*, 225(6):441–448, 2011.

[154] M. H. Müser, W. B. Dapp, R. Bugnicourt, P. Sainsot, N. Lesaffre, T. A. Lubrecht, B. N. J. Persson, K. Harris, A. Bennett, K. Schulze, et al. Meeting the contact-mechanics challenge. *Tribology Letters*, 65(4):118, 2017.

[155] C. Wen, X. Meng, and W. Li. Numerical analysis of textured piston compression ring conjunction using two-dimensional-computational fluid dynamics and reynolds methods. *Proceedings of the Institution of Mechanical Engineers, Part J: Journal of Engineering Tribology*, page 1350650118755248, 2018.

Journal Publications

[CFM+18] A. Codrignani, B. Frohnapfel, F. Magagnato, P. Schreiber, J. Schneider, and P. Gumbsch. Numerical and experimental investigation of texture shape and position in the macroscopic contact. *Tribology International*, 122:46–57, 2018.

[CSMF17] A. Codrignani, D. Savio, F. Magagnato, and B. Frohnapfel. A scaling parameter for pressure losses and thermal effects in lubricant flows with viscous dissipation. *Tribology International*, 113:238–244, 2017.

[GMB+15] C. Greiner, T. Merz, D. Braun, A. Codrignani, and F. Magagnato. Optimum dimple diameter for friction reduction with laser surface texturing: the effect of velocity gradient. *Surface Topography: Metrology and Properties*, 3(4):044001, 2015.

Co-Supervised Student Theses

[Alb17] Linda Albert. Implementation of roughness models in the Reynolds equation. *Bachelor thesis at Karlsruhe Institute of Technology*, 2017.

[Fur16] Oscar Furst. Analysis of 3D surface textures with the reynolds equation. *Bachelor thesis at Karlsruhe Institute of Technology*, 2016.

[GB17] Benjamin Guthier-Brown. Numerical analysis of surface textures for lubricant applications. *Bachelor thesis at Karlsruhe Institute of Technology*, 2017.

[Lop17] Eduardo Lázaro Lopez. CFD analysis of heat transfer in microchannels. *Master thesis at Karlsruhe Institute of Technology*, 2017.

[Ros16] Tommaso Rossi. Numerical analysis of surface pattern for a clutch. *Master thesis at Karlsruhe Institute of Technology*, 2016.

[Shu16] Jingzhe Shu. Investigation of inertia effects in lubricant flows with textured surfaces. *Bachelor thesis at Karlsruhe Institute of Technology*, 2016.

[Ukr18] Ihor Ukrainets. Numerical modeling of hydrodynamic effects in mixed lubrication regime. *Bachelor thesis at Karlsruhe Institute of Technology*, 2018.

Nomenclature

TRIBOMETER	
R_a	arithmetic mean roughness of a surface
R_f	friction radius
F_N	normal force
F_T	tangential force
C_f	friction coefficient x
R_{pin}	pin radius
h_{side}	height of the pin (from the edge to the pin holder)
s_h	shoulder height of the pin
$\mathcal{T}$	torque applied on the tribometer shaft

SIMULATIONS	
x	streamwise direction
z	spanwise direction
y	wall normal direction
L_x	length of the domain
L_z	width of the domain
H	gap height
p	pressure

p_{cav}	cavitation pressure
$\mathbf{p}$	discretized pressure vector
ρ	density
ρ_{ref}	reference density
$\bar{\rho}$	generic density as constant value
T	temperature
θ	cavity fraction
$\boldsymbol{\theta}$	discretized cavity fraction vector
N_x	amount of grid nodes in streamwise direction
N_y	amount of grid nodes in wall-normal direction
N_z	amount of grid nodes in spanwise direction
N	total amount of grid nodes
k_T	temperature diffusion coefficient
$\mathbf{J}$	Jacobian matrix
$\mathbf{J}_r$	preconditioned Jacobian matrix
$\mathbf{p}$	discretized pressure vector
$\boldsymbol{\theta}$	discretized cavity fraction vector
$\mathbf{F}$	Fischer-Burmeister matrix
$\mathbf{G}$	matrix representation of the Reynolds equation
ϵ_T	pressure residue
ϵ_p	temperature residue
H^t	total enthalpy
$\mathbf{V}$	Volume of the channel

Non dimensional number

S	Sommerfeld number
Re	Reynolds number
Pe	Péclet number

Lubricant

μ	dynamic viscosity
ν	kinematic viscosity
ρ_{ref}	reference density
μ_0	reference dynamic viscosity
C_2	exponential coefficient for the viscosity-temperature relation
T_0	reference temperature

Texture

ρ_{txt}	texture density
D	dimple diameter
De	dimple depth
α	angle between two lines of the dimple distribution
l_x	streamwise distance between two adjacent dimple
l_z	spanwise distance between two adjacent dimple
s_z	spanwise shift between two consecutive lines of dimples

ABBREVIATIONS	
DNS	direct numerical simulation
EHL	elasto-hydrodynamic lubrication
TEHL	thermo-elasto-hydrodynamic lubrication
FBNS	Fischer-Burmeister-Newton-Schur (Cavitation algorithm)
LST	laser surface texturing
AFM	atomic force microscopy
PAO	Poly-alpha-olefin (lubricant oil)
FVA	Forschungsvereinigung Antriebstechnik (german consortium for lubricant oils)
EOS	equation of state
MPI	message passing interface
ISTM	Institut für Strömungsmechanik
KIT	Karlsruher Institut für Technologie

List of Figures

List of Tables

A Modelling of the lubricant properties

In this appendix, the rheological properties of the lubricant oil, which was considered in the numerical simulations, are presented. The rheological description of a fluid concerns two main steps, namely the representation of the thermodynamic behaviour through an equation of state in the form of $\rho(T,p)$ and the choice of a constitutive law for the characterization of the viscosity as a function of the thermodynamic state $\mu(T,p)$.

Among the four different experimental campaigns which are presented in section 2.1, three different kinds of lubricant oil were employed. Moreover, in the analysis of the effects of viscous dissipation presented in chapter 3, a fourth kind of lubricant oil was considered. This variety in the choice of lubricants were imposed by different necessities in the carrying out of the experiments. Nonetheless, in the numerical representation of the cited experimental campaigns, particular attention is paid in the faithful description of the properties of the lubricant and for every analysis the parameters of the oil are change in order to match the physical properties of the oil used in the experiments.

For this reason, the following equations of state and the constitutive laws were used for different lubricant parameters. However, for sake of brevity, the graphs in the following discussion will be presented for the case of the mineral oil considered for the viscous dissipation analysis in chapter 3. All other oil parameters used for the other analysis are just reported in the table A.1.

Equation of state for lubricant oils

Two equations of state (EOS) are analyzed in this work, namely the stiffened gas equation of state (SG) and the Dowson-Higginson (DH). Both of them are considered in the form $\rho(T,p)$, where the density is expressed as function of the temperature and pressure. This formulation is typical for the solution of fluid dynamic problems, since pressure, temperature and velocity fields are commonly computed through the solution of the mass, momentum and energy equations, leaving the computation of the density as an iterative update at the end of every iteration (as shown, for example, in the algorithm flow chart in figure 2.5).

The stiffened gas equation of state is based on a reformulation of the ideal gas EOS applied for compressible liquids [111]. This equation of state is typically expressed in the following formulation:

$$\rho(T,p) = \frac{p + \gamma_{sg} P_{sg}}{R_{sg} T} \tag{A.1}$$

where γ_{sg} and P_{sg} are two constants which depend on the characteristics of the fluid and are tuned in order to fit the experimental measurements. The stiffened gas constant R_{sg} is defined as $R_{sg} = C_v(\gamma_{sg} - 1)$, where C_v is the heat capacity at constant volume.

The lubricant chosen for the study in chapter 3 is a mineral base oil whose properties were measured experimentally by Regueira *et al.* in [110]. Such kind of oil is chosen for the twofold reason of being commonly used in many tribological applications and also because of the availability of experimental works which provide a detailed experimental characterization of it, such as the one by Regueira *et al.* [110].

k	0.154 [W/(mK)]
γ_{sg}	4.0606
P_{sg}	$4.27871 \cdot 10^8$ [Pa]
C_v	2000 [J/(KgK)]
A_1	$-7.2447 \cdot 10^{-4}$ [1/K]
A_2	$2.2444 \cdot 10^{-7}$ [1/K^2]
B_1	$6.6009 \cdot 10^{-10}$ [1/Pa]
B_2	$2.8225 \cdot 10^{-9}$ [1/Pa]
ρ_0	877.7007 [Kg/m^3]
P_0	101325 [Pa]
T_0	303.1500 [K]
μ_0	$1.871 \cdot 10^{-2}$ [Pa·s]
c_1	-0.070712 [1/K]
c_2	313.0 [K]

Table A.1: Rheological and thermal parameters of the lubricant, considered for the characterization of the lubricant oil in [CSMF17].

The second equation of state which was employed is the Dowson-Higginson one [58]. This is one of the most used EOS in tribology, because of its ability to fit experimental data without increasing the complexity of model excessively. This work considers the DH EOS in the formulation presented in [110]:

$$\rho(P,T) = \rho_0 \left(1 + A_1(T - T_0) + A_2(T - T_0)^2\right) \left(1 + \frac{B_1(P - P_0)}{1 + B_2(P - P_0)}\right). \quad \text{(A.2)}$$

The corresponding coefficients are summarized in Table A.1 along with thermal and material properties of the fluid.

Figures A.1 and A.2 show the dependence of the pressure on density at two different temperature values as portrayed by the stiffened gas EOS and the Dowson-Higginson EOS, respectively. Both figures show also the comparison

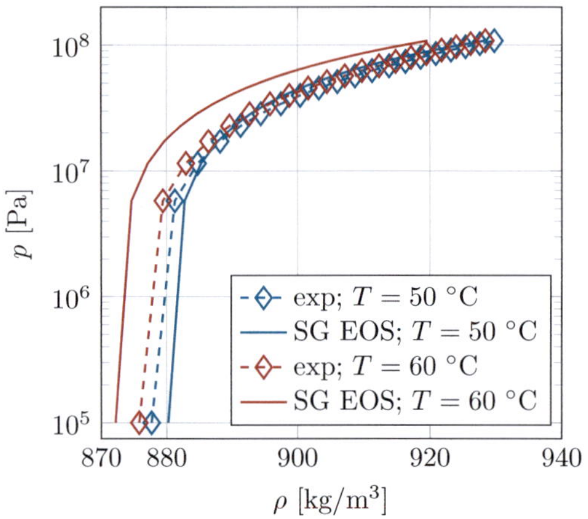

Figure A.1: Stiffened gas equation of state for the mineral base oil considered in the study of the viscous dissipation effects presented in chapter 3. This graphs represent the relation $p(\rho,T)$ by showing the dependence of the pressure on the density for two values of temperature. The experimental data are taken from the work of Regueira *et al.* [110].

with the experimental data from [110]. As on can see, the Dowson-Higginson model provides a better match of the experimental data than the stiffened gas one. This is also due to the fact that the Dowson-Higginson EOS has more degrees of freedom which allow to find a better fit of the data, in particular for what concerns the temperature variations. The optimal fit is found in the present work through the application of a genetic algorithm.

Constitutive laws

The analyzed lubricant is always assumed to be Newtonian in the present work. This hypothesis is verified to hold for great part of the numerical work performed in the hydrodynamic region in chapters 3 and 4, where the maximum shear rate $\dot{\gamma} = 10^6$ is reached only for a very small portion of the investigated parameter space.

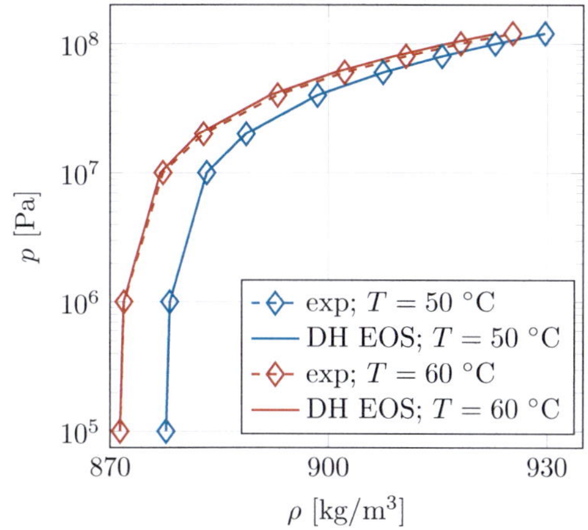

Figure A.2: Dowson-Higginson equation of state for the mineral base oil considered in the study of the viscous dissipation effects presented in chapter 3. This graphs represent the relation $p(\rho,T)$ by showing the dependence of the pressure on the density for two values of temperature. The experimental data are taken from the work of Regueira *et al.* [110].

The viscosity is assumed to vary exponentially with the temperature and since the pressure variations remain always relatively small, no dependence on the pressure in the viscous law is considered in this work.

$$\mu(T) = \mu_0 e^{c_1(T-c_2)} \tag{A.3}$$

The three coefficients are calibrated through the experimental data by Braun *et al.* in [19]. Figure A.3 shows the exponential relationship exposed in equation A.3 together with the experimental measurements of the viscosity. In the parametric study in chapter 4 the viscosity is varied by keeping as reference the viscosity value $\mu_0 = 1.871 \cdot 10^{-2}$ at $T = 30$ °C.

Internal energy depends linearly on the temperature follows the definition $E_i(\rho,T) = \rho C_v T$ (where $E_i(T=0) = 0$) with the assumption of constant specific heat coefficient at constant volume, $C_v = const.$

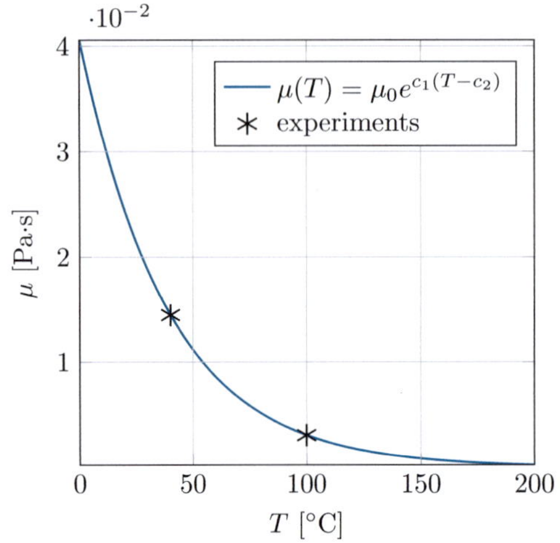

Figure A.3: Constitutive law of the PAO-18 lubricant. The graph shows the dependence of the viscosity on the temperature which is modeled through the exponential laws shown in equation A.3. The coefficients μ_0, c_1 and c_2 are shown in table A.1. The two points represent the experimental measurements as provided in the work of Braun *et al.* [19].

B Role of the texture density in pin-on-disc set-ups

In this appendix, the analysis on influence of the textured area is presented starting from the experimental campaign carried out by Schneider *et al.* in [20] by means of the tribometer set-up introduced in section 2.1. Figure B.1 shows the experimental Stribeck curve with different values of textured area ranging between 5% and 30%. The dimple diameter is kept constant to 40 μm with a constant aspect ratio $\lambda = 10$. As one can see, texturing the 10% of the surface area brings the best results in terms of friction reduction. As said in the introduction to this chapter, this findings obtained by means of a tribometer set-up are in agreement with previous literature results presented, for example, by Klingerman *et al.* in [117] for piston-rig systems and by Shinkarenko *et al.* in [119] for seals.

In the corresponding numerical analysis the same texture configuration is analyzed and, moreover, particular attention is paid to the role of the viscosity on the definition of the optimal texture density. The main results of the simulations are shown in figure B.2 where the normalized friction coefficient $\bar{C_f}$ is plotted against the texture density ρ_{txt}. the normalized friction coefficient is defined as the ratio between the friction coefficient in the texture case $C_{f_{txt}}$ and the reference one in the untextured case C_{f_0} so that $\bar{C_f} = C_{f_{txt}}/C_{f_0}$. It is interesting to notice that the friction normalized friction coefficient decreases monotonically with increasing texture density. On the contrary, experimental investigation have proved that an optimal texture density exists around $\rho_{txt} = 10\%$. An explanation of this discrepancy was recently proposed by Wang *et al.* in [127] by analysis a single texture element with an approach

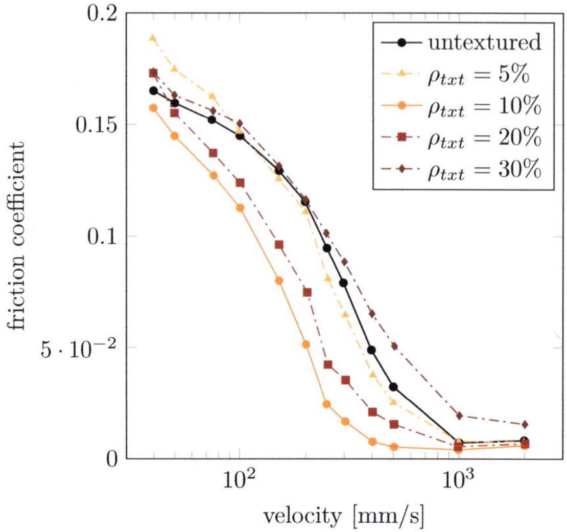

Figure B.1: Experimental Stribeck curve as function of the sliding speed for different values of the texture density ρ_{txt} as reported by Schneider *et al.* in [20]. The dimple diameter and depth are kept constant to $D = 40$ μm and $h_D = 4$ μm respectively. The texture pattern is hexagonal, further details about the experimental set-up are reported in section 2.1.

based on the contact mechanics. That study concluded that a pure fluid dynamic approach cannot take into account the increasing forces exchanged through the direct contact of the two surfaces as soon as the untextured area reduces due to an excessive presence of texture elements.

Beside confirming the literature results, the current study aims also to study the influence of the operating condition on the texture design parameters. For this reason, five values of viscosity are shown in figure B.2 ranging from the reference value $\mu_0 = 0.0187$ Pa·s to a value which is 20 times higher: $\mu = 0.374$ Pa·s. As one can see, the increase of the viscosity makes the effects of texture density more effective, since a much lower normalized friction coefficient can be achieved for higher values of viscosity at the same level of textured area. This could mean that also the optimal texture density value could be actually very sensitive on the viscosity and this could bring to the identification of a

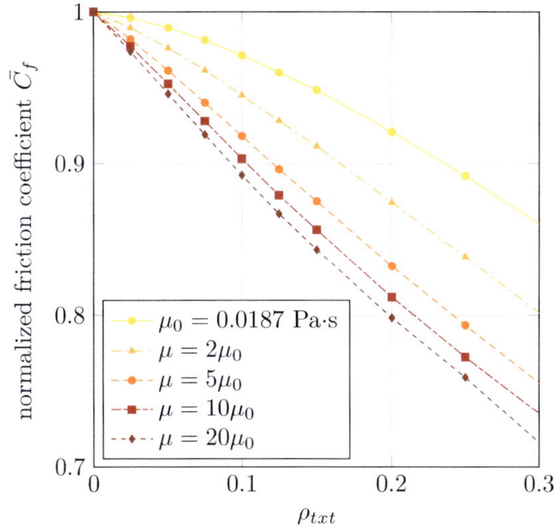

Figure B.2: Normalized friction coefficient $\bar{C_f}$ as a function of the textured density ρ_{txt}. The normalized friction coefficient is computed as the ratio between the friction coefficient in the texture case and the reference one in the untextured case. The dimple diameter and depth are kept constant to $D = 40$ μm and $h_D = 4$ μm respectively as in the corresponding experimental analysis by Schneider *et al.* [20].

different optimal ρ_{txt} in comparison to what already done in literature [117–119]. Further study should also consider the role of the contact mechanics as underlined by Wang *et al.* in [127].

Acknowledgements

Embarking on a PhD is for sure a challenging choice, whose achievement would have not been possible without the help and the constant support of long list of people, whom I really own much.

First of all, I want to thank my doctoral adviser Prof. Dr.-Ing. Bettina Frohnapfel. Her careful and sensible guidance was a key factor in the attainment of this doctorate and in my professional development. Thank you for having accepted me as PhD student and for giving me the chance to work in a very motivating and friendly environment.

Further, I want to express my deep gratitude to Prof. Maurizio Quadrio, who constantly advised and encouraged me in the last years, especially during the turbulent times. Thank you for having made this experience possible.

A special word of thanks is owned to Dr. Daniele Savio, who never hesitated to share with me his knowledge in the field of tribology, giving a crucial contribution to the development of this work.

This work has also thoroughly benefited from the suggestions and inputs from Prof. Dr. rer. nat. Peter Gumbsch, who put the basis for a florid collaboration between our institute and the institute of applied materials. In this regards I would like to thank also Dr.-Ing. Johannes Schneider, M. Sc. Paul Schreiber for the pleasant and fruitful discussions during our efforts to build a bridge between experiments and simulations.

I would like to express my gratitude to the Friedrich und Elisabeth Boysen Stiftung für Forschung und Innovation for the financial support during the last years.

I would also like to thank the students, who did their thesis under my guidance for their commitment and their curiosity, which definitely improved the quality of my research and also motivated me much.

I am deeply grateful to my whole institute for the wonderful working environment and all the time that we enjoyed together also outside the campus. Alexander, Andreas, Anna, Davide, Frieder, Giacomo, Jochen, Karin, Kay, Lars, Pourya, Sibela, Steffen, Tobias and Verena, without doubt, one of the biggest gift of this experience was the chance to meet you and share these years with you. I hope we will always go on with our epic trips!

A huge thanks goes to all the people who shared with me the passion for music and theater: the wonderful members of Unitheater, SAX crippled Thumbs, Paradiso and also Stefan, Niko, Philipp and Peter. A particular mention deserves my beloved band String Theory, for their love and patience which were never affected by the distance. Thank you all for having made this journey into the scientific world more colorful and tasty through your art and music!

Finally, I would like to thank also my lifetime friends who have always been there despite the distance. Brenda, Andrea, Marco, Stefano and all my former colleagues from the Polimi. A huge hug goes to Paola and Francesco for their never-stopping life mentorship.

Tutto ciò non sarebbe stato possibile senza il grande e infinito supporto della mia famiglia, mamma, papà e Davide. Grazie per aver appoggiato senza esitazione la decisione di imbarcarsi in questa grande avventura e soprattutto grazie per aver instaurato in me fin da piccolo la curiosità e la passione per la scoperta del mondo.

Schriftenreihe des Instituts für Strömungsmechanik

Karlsruher Institut für Technologie (KIT) | ISSN 2199-8868

Band 1 **Gertraud Maria Daschiel**
Strategies to reduce friction losses and their implications for the energy efficient design of internal flow domains.
ISBN 978-3-7315-0291-3

Band 2 **Andreas Güttler**
High accuracy determination of skin friction differences in an air channel flow based on pressure drop measurements.
ISBN 978-3-7315-0502-0

Band 3 **Anna Christina Daub**
Numerical Haemodynamics in the Human Heart.
ISBN 978-3-7315-0757-4

Band 4 **Alexander Stroh**
Reactive Control of Turbulent Wall-Bounded Flows for Skin Friction Drag Reduction.
ISBN 978-3-7315-0766-6

Band 5 **Giacomo Cocconi**
Numerical Study on the Propagation of Turbulent Fronts in Dilute Polymer Solutions.
ISBN 978-3-7315-0829-8

Band 6 **Andrea Roberto Codrignani**
Numerical representation of a pin-on-disc tribometer for the investigation of textured surfaces.
ISBN 978-3-7315-0874-8